Battledrills for Chinese Mobile Warfare

OTHER TACTICS/INTEL. MANUALS FROM POSTERITY PRESS:

ADVANCED TACTICS IN AMERICA
OKINAWAN RECKONING
PELELIU PROGRESS
THE IWO ALAMO
SUPER-SQUAD: THE NOW MISSING COMPONENT
NORTH KOREA'S HIDDEN ASSETS
GUARDIAN JOE: HOW LESS FORCE HELPS THE WARRIOR
SINOLAND: THE SUBVERSION OF FREEDOM'S BASTION
AFRIQUE: A WARNING FOR AMERICA
STRATEGIC RIFLEMAN: KEY TO MORE MORAL WARFARE
GUNG HO: THE CORPS' MOST PROGRESSIVE TRADITION
GLOBAL WARRIOR: AVERTING WWIII
EXPEDITIONARY EAGLES: OUTMANEUVERING THE TALIBAN
HOMELAND SIEGE: TACTICS FOR POLICE AND MILITARY
TEQUILA JUNCTION: 4TH-GENERATION COUNTERINSURGENCY
DRAGON DAYS: TIME FOR "UNCONVENTIONAL" TACTICS
TERRORIST TRAIL: BACKTRACKING THE FOREIGN FIGHTER
MILITANT TRICKS: BATTLEFIELD RUSES OF THE ISLAMIC MILITANT
TACTICS OF THE CRESCENT MOON: MILITANT MUSLIM COMBAT METHODS
THE TIGER'S WAY: A U.S. PRIVATE'S BEST CHANCE FOR SURVIVAL
PHANTOM SOLDIER: THE ENEMY'S ANSWER TO U.S. FIREPOWER
ONE MORE BRIDGE TO CROSS: LOWERING THE COST OF WAR
THE LAST HUNDRED YARDS: THE NCO'S CONTRIBUTION TO WARFARE

Battledrills for Chinese Mobile Warfare

Illustrated Edition

H. John Poole
Foreword by
Col. Eric M. Walters USMC (Ret.)

Posterity
Press®

Published by Posterity Press
P.O. Box 5360, Emerald Isle, NC 28594
(www.posteritypress.org)

Cataloging-in-Publication Data
Poole, H. John, 1943-
Battledrills for Chinese Mobile Warfare
Includes bibliography and index.
1. Infantry drill and tactics.
2. Military art and science.
3. Military history.
I. Title. ISBN: 978-1-7354530-3-3 2023 355'.42
Library of Congress Control Number: 2023920156

Edited by Dr. Mary Beth Poole

First printing, United States of America, November 2023

For all U.S. military personnel who realize that it takes dismounted small-unit infantry skill to decisively win a war.

CONTENTS

Illustrations

Figures:

TABLES

Foreword

> "Too much of what we consider 'normal' in military tactics is just blind obedience to long-past practices that have become ingrained in our current thought. Question everything (Oliviero, *Praxis Tacticum,* Double Dagger Books, 2021, vii)."
>
> — Col. (Ret.) Charles S. Oliviero, CD, Ph.D.

John Poole has cemented his reputation for questioning everything having to do with U.S. small infantry unit tactics over several decades, as evidenced by the majority of his 23 books published before this one. He questioned our traditional tactics, techniques, and procedures through the eyes of our enemies in past history, in current contemporary circumstances, and in hypothesized future conditions. His microscopic dissections of past evidence reflect his dedication toward the "highest use of history to teach wisdom from the mistakes of others."

Often judged by some with military experience as controversial, his well-researched and reasoned challenges to established tactical methods are not so easily dismissed, even if they appear counter-intuitive at first blush.

This latest work of his is no exception, tailored to what we might expect our potential People's Liberation Army ground force small unit adversaries to perceive and act when engaged in close quarters battle with us. Those familiar with Poole's work will immediately recognize the distillation of his earlier assessments and advice in this volume. For new readers introduced to the author for the first time, *Battledrills for Chinese Mobile Warfare* serves as both a warning and a guide to practical action. Indeed, the latter is Poole's emphasis here, intended for infantry leaders—junior and senior non-commissioned officers in particular—to practice both in garrison and out in the field. No method or technique can be evaluated apart from its actual execution, repeated in different free-play scenarios, and subject to rigorous collective self-critique. In this way, the unit not only teaches itself to be proficient in a particular skill, but also to recognize when a change in execution is imperative and to adapt accordingly.

Senior leaders also have much to gain from this book. Small unit disciplined initiative can only prosper and develop given the right command climate. That is indispensable. The author of *Praxis Tacticum: The Art, Science, and Practice of Military Tactics,* Col. Chuck Oliviero, advises:

> Everywhere, military commanders claim that of course they allow subordinates to learn through mistakes; the truth is somewhat different. Whatever the case, if your soldiers do not feel confident that they can experiment and learn as they practice new or different tactical solutions, they will revert to safe—even if unsuccessful—solutions (Oliviero, *Praxis Tacticum,* Double Dagger Books, 2021, 79).

Poole's enduring mission is to insist that no solution imposed from high above—especially when facing unexpected challenges in contact with the enemy—can succeed as well as one a self-taught and well-practiced infantry small unit can confidently improvise on the spot. This has implications both for ground force unit headquarters and for the formal schools and operating force exercise and evaluation organizations that train individuals and these small units.

There was a time when the Department of Defense appeared to recognize this, initiating the "Close Combat Lethality Task Force" (CCLTF) that was less focused on technological solutions and concentrated more on the human dimension of combat proficiency. While the CCLTF still exists as of 2023 at Fort Moore, Georgia, it remains an open question whether or not the primacy of human factors in improving tactics has yet been adequately infused into the general purpose Army and Marine Corps infantry the CCLTF is meant to serve.

If one doubts that American infantry has escaped its industrial age legacies that Poole describes in his books, self-education is one way forward. But such self-education cannot be limited by emerging doctrine alone. As J.F.C. Fuller argued:

> [T]oday we are faced by so rapid a development, or evolution, in administration, strategy, and tactics, and through these in organization, command and discipline, that this development constitutes a revolution which renders our existing art of war obsolete, so obsolete, that unless we can grasp what it portends, to rely on it in another war is likely to prove a greater danger than to enter it totally ignorant of military values. The reason for this is, that an open mind will readily adapt itself to circumstances, whilst one crowded with obsolete knowledge must first disencumber itself before it can adapt itself to surroundings (Maj.Gen. J.F.C. Fuller, *Lectures on F.S.R. III,* London: Sifton Praed & Co, 1932, 2).

Battledrills for Chinese Mobile Warfare, like Poole's earlier works, goes beyond doctrinal processes. It provides an essential stepping-stone on the path of professional self-development through not just study with the open mind of an individual, but also collective practice within the unit.

Col. Eric M. Walters USMC (Ret.)
former assistant professor,
U.S. Army Command & Gen. Staff College

INTRODUCTION

America's traditional adversaries are from bottom-up Eastern societies that better understand how to limit frontline casualties. For such armies, manpower-efficient small-unit tactics drive the operations, and successful operations shape the strategies. But, the Pentagon still gets its strategies from people too talented to have filled any combat job for long. So, as World War III (WWIII) looms, all United States (U.S.) infantry procedures must be reassessed from the standpoint of common sense. U.S. government leaders have no shortage of answers, but may have skipped a few questions.

The differences in foreign army semantics can be revealing. Most Western nations now limit the number of "ground pounders" they commit to any conflict. Nor can their forces rapidly restore peace to a region. The reason is clear from the People's Republic of China's (PRC's) choice of terms. What Americans call "Attrition Warfare" is "Positional Warfare" to the Communist Chinese. Their grass-roots perspective easily identifies the Yank way of war as too plodding and destructive—of being more defensive than offensive in nature. That's why Beijing also requires a Mobile Warfare capability of every size unit (where the term implies no mechanical conveyance whatsoever). While similar to the Maneuver Warfare (MW) of the WWI Germans, it predates MW by thousands of years. Among its additional parameters are how quickly to disperse and reassemble a big foot infantry outfit. To a Western observer, such a thing looks a lot like Guerrilla Warfare—the third Chinese prerequisite.

> As defined by Giap, "[M]obile warfare is the fighting way of concentrating troops of the regular army in which relatively big forces are regrouped and operating on a relatively vast battlefield, attacking the enemy where he is relatively exposed . . . , advancing very deeply and then withdrawing

> very swiftly" (Giap, *People's War—People's Army,* 1962, 106). In essence, mobile warfare involved large units conducting operations against the enemy, but it did not direct fighting set-piece battles. . . . Put simply, mobile warfare and guerrilla [warfare] were inseparable and mutually supporting (Op. Cit., 1962, 107-108).[2]
>
> — *The Battle of Ap Bac, Vietnam,* by Toczek

While both U.S. infantry branches have claimed something like MW as their doctrine since the late 1980's, overly tall organizational structures have kept them from fully practicing it. That would take decentralization of control—the antithesis of top-down society. Nor has either branch tried to field maneuver elements small/agile enough to escape modern-day targeting devices. To complete an MW mission, such an element must sneak into an enemy strongpoint or bypass it. The latter takes speed. The standard U.S. patrol is too slow and deliberate. Proficiency at MW takes constant refinement to one's dismounted movement technique.

> If the tiger stands his ground, the elephant will crush him with his mass. But, if he conserves his mobility, he will finally vanquish the elephant, who bleeds from a multitude of cuts.[1]
>
> — Gen. Vo Nguyen Giap

Chinese Mobile Warfare is neither mechanized nor motorized. It consists of tiny infantry increments constantly in transit to escape Western artillery, tank, and aircraft strikes. America was able to defeat its Japanese version on isolated islands in World War II (WWII), but not without excess casualties. A higher tempo of operations against the North Vietnamese Army (NVA) variant failed miserably in Vietnam. Both times, more U.S. small-unit skill and initiative had been required.

Much of what U.S. warriors still believe to be combat-worthy method may not be. It is the product of an Attrition Warfare mindset. A student of Chinese Mobile Warfare might realize another option. Both would work, but the latter result in fewer losses.

LT.COL. H. JOHN POOLE USMC (RET.)
CAREER INFANTRYMAN AND
FORMER FLEET MARINE FORCE GY.SGT.

ACKNOWLEDGMENTS

Once again, God gets the credit for anything useful from this latest research. Tiny-element techniques will continue to be the key to decisively winning a war with minimal casualties. According to Pope John Paul II, the most moral form of combat causes the least damage to, and possibly saves the life of, the opponent.[1] That requires more "moving" than "shooting." This work will attempt to "communicate" the particulars.

PART ONE

THE ONGOING NATURE OF GROUND COMBAT

"MY MANY YEARS IN THE ARMY HAVE DEMONSTRATED THAT WHEREVER CONFIDENCE IN NCOs IS LACKING . . . YOU HAVE . . . NO REALLY COMBAT-WORTHY UNITS."
— GEORGI K. ZHUKOV, VICTOR OF STALINGRAD AND BERLIN

(Source: Georgi K. Zhukov, *Reminiscences and Reflections,* 1974, in *Warriors' Words,* by Peter G. Tsouras [London: Cassel Arms & Armour, 1992], p. 282.)

1 THE MODERN-DAY MYTH

- Why is America now so hesitant to put "boots on the ground"?
- Does it have anything to do with their degree of skill?

National agencies have no chronic deficiencies.

(Source: Free clipart courtesy of pixy.org, image designator "pixy.org/963190/.")

With Respect to the Current Generation

On 1 May 2023, a TikTok video hit the U.S. internet entitled "The Camels Are on the Horizon." While TikTok's sponsor only has camels in its breakaway Moslem provinces, it has always been good at shifting the blame for unseemly deeds. Ever since 9/11, it has been doing everything possible to undermine America's internal stability.[1] In late March of 2023, the Border Patrol reported the number of Chinese nationals entering the U.S. had just spiked by 1,000%.[2]

That Asian video talks of empires self-destructing after 250 years through too much soft living. But, America is not an empire; it's an aspiration and an experiment. It has tried hard over the years to assist (rather than dominate) other nations. And, its current generation is no less resilient than those before it, just a little different.

Being out of shape and too heavily dependent on one's cell phone are not irreversible flaws in a prospective warrior. "Free spirits" had won the Air Battle of Britain in 1940. And, young Americans of today have many unique strengths. Unlike their fathers and grandfathers, they have had to excel at several different jobs and even professions. They also possess more affinity for urban combat. To escape the increasingly lethal ordnance, most future wars will be fought in heavily populated areas.

So, that foreign video could serve a useful purpose. Bottom-up societies more easily produce tiny, self-sufficient infantry elements—those not needing a supporting-arms umbrella. America may soon find itself in a shooting war with TikTok's sponsor. To win that war, U.S. units must acquire a more Oriental way of fighting.

Good Intentions Aren't Always Enough

American parents love both country and offspring. But, they have trusted a top-down bureaucracy with the battlefield welfare of those offspring. And, doing exactly as told won't get the best of an initiative-guided adversary. So, the U.S. defense establishment tries to win all foreign conflicts through equipment alone. It would take hundreds of semi-autonomous infantry squads to consolidate any large attack objective. That's why no U.S. expeditionary force has decisively won a major ground war since WWII.

The Current Emphasis on Virtual Reality

U.S. infantry tactics have become so standardized as to be totally predictable in wartime. Supposedly kept simplistic so that GIs (government issues) can more easily adapt them to current circumstances, they have instead become what bone-weary soldiers do without any instructions to the contrary. Disregarding one's surroundings in combat has dire consequences.

When first written, the manuals contained logical steps for various small-unit missions. But, many of those steps were too blatantly apparent to the enemy, and their parent maneuver still pre-machinegun (MG) in format. A foe automatic weapon is most safely closed upon from a flank through the microterrain by a prone individual. A good MG bunker will withstand direct heavy caliber hits. So, only a grenade through its gun port can reliably silence it for a while. The "crawling beneath strand barbed wire and frontal MG fire" that U.S. infantrymen used to experience in training fails to emphasize that indirect, covered approach. And, most GIs still don't know how to best deal with an active gun. With more dispersion now dictated by advances in enemy weaponry, such an oversight can no longer be tolerated by the Pentagon. Too much standardization has become lethal. And, the close-quarters (built-up) nature of future battlefields will no longer permit large-unit formations. That means each rifle company can have its own squad stratagems without adversely affecting the whole.

Mobile Warfare Allows for Fewer Friendly Casualties

Most wars are won through offensive action. Surprise and firepower are now widely known to be interchangeable.[3] A fully befuddled defender can't hurt his attacker. That's why assault troops prefer one of the 3rd Generation Warfare (3GW) variants. Mobile Warfare is a more sophisticated Asian version of the MW developed by the World War I (WWI) Germans and adopted by the United States Marine Corps (USMC) around 1988.

The central tenets of 3GW are often diametrically opposed to those of the "higher-tech" Attrition Warfare still practiced by most American infantry outfits. (See Tables 1.1.) As a result, 3GW's control parameters may seem overly loose to fully indoctrinated GIs. (See Table 1.2.)

While both 3GW versions call for semi-autonomous small units, "advanced" Attrition Warfare too tightly monitors the weapons safety of (micromanages) small units. So, more 3GW must be practiced by Freedom's Bastion (along with its traditional way of war) to successfully conclude the next global conflagration. Only Mobile Warfare allows for the dispersion of any unit big enough to attract a drone strike. Thus, more familiarity with that Oriental process is well worth the effort.

Maneuver Warfare	Attrition Warfare
Tries to Bypass or Demoralize Foe	Kills as Many Foe as It Can
Creates as Much Surprise as Possible	Depends on Fire Superiority
Focuses Outward on Enemy	Fixated with Own Conduct
Uses Reconnaissance Pull	Relies on Command Push
Moves Through Undefended Gaps	Hits Enemy Concentrations
Exploits Breakthroughs	Maintains Unit Alignment
Captures Centers of Gravity	Seizes Hilltops
Employs Combined Arms	Uses Biggest Gun Available
Mostly Operates at Night	Prefers Day Attacks
Prefers Non-Illuminated Maneuvers	Assaults Under Illumination
Seeks High Tempo of Operations	Is Slow and Methodical
Defends Only as a Trap	Holds Its Position Often
Sometimes Pulls Back	Never Retreats
Routinely Delegates Authority	Closely Follows Headquarters
Offensively Oriented	Largely Defensive in Nature

Table 1.1: The Two Main Styles of War Are Veritable Opposites
(Source: *Table 2.1 of "The Tiger's Way," by Poole.*)

Maneuver Style	Attrition Style
Initiative Encouraged	Everyone Does as Told
Subordinates Keep Boss Informed	Alternative Plan Takes Permission
Mission Type Orders	Detailed Orders
Control Via Rehearsed Methods	Control Through Leader Directives
Sub-Unit Techniques Authorized	All Methods Standardized
Trains to the Threat	Prepares as Usual
Not Much Signaling Required	Complicated Communications
Decentralizes Control	Lower Echelons Micro-Managed
More Gung Ho Squad Chemistry	Susceptible to Bureaucratic Inertia

Table 1.2: They Have Vastly Different Control Parameters

Aren't Two Choices Better Than One?

Because the characteristics of 2nd Generation Warfare (2GW) and 3GW are so markedly different, the Pentagon attempted to replace one with the other in the late 1980's. That effort failed because the doctrinal change had not been accompanied by commensurate small-unit methods. And, many U.S. leaders were too proud to admit that their traditional ways of doing things weren't always the best. Instead of 2nd Amendment Rights, a better understanding of when to opt for surprise over shooting would have done them more good. (Look again at Tables 1.1 and 1.2.) Like the outnumbered Asian citizen, a U.S. rifleman would welcome unusual ways of doing things.

2 How Much Has the New Gear Changed Things?

- Why hasn't warfare changed much since advent of the MG?
- Which countries still play down its existence?

What's different for the U.S. rifleman?

(Source: http://search.usa.gov public-domain image from this url: http://usarmy.vo.llnwd.net/e2/-images/2006/11/21/990/size0-army.mil-2006-11-21-102805.jpg)

How Has the Latest Ordnance Affected the War in Ukraine?

America's military magazines are full of ads for new infantry paraphernalia. U.S. troops spend many hours each week learning how to operate the stuff. And, the American people are constantly told their service branches are the best in the world at all things. Yet, the war in Ukraine has now regressed into an old-fashioned slugfest. Might the equipment being used have less to do with its progress, than the tactics?

After the Kyiv-bound Russian column met the same fate as the one entering Grozny in 1994, the liberation of Ukraine has sadly slowed to a crawl. A Finnish Army observer believes it will eventually fail through too little squad training.[1] The Finns have firsthand experience with Russian invasions. Motti Tactics in 1939 permitted them to concede only the tiny Karelian Isthmus, some land north of Lake Ladoga, and a small part of the northern Salla region to the powerful Soviets as part of the Moscow Peace Treaty of March 1940.[2]

The stopping of the Kyiv convoy had been facilitated by Javelin antitank missiles and the lack of city defender organization. As at Grozny, a swarm of tiny units had been too much for the stalled column of vehicles to endure.

Then, U.S. assistance turned to artillery. A standoff approach to war took the place of up-close-and-personal contact. Before the U.S. Congress, President Zelenskyy admitted that all that artillery wasn't doing his country much good. In effect, the war in Ukraine was quickly becoming another trenchline nightmare from WWI. To win it, the Ukrainians would need hundreds of modern tanks, air cover, and large numbers of highly disciplined troops. But, according to that Finnish report, Ukrainian soldiers lacked squad structure and training, and instead lounged around their barracks waiting for the next assignment.[3]

So, the only thing really different about the Ukrainian War is an abundance of drones. They provide more advance notice of approaching infantry formations and armor than before. But, that had already been possible through satellite imagery, just not as quickly actionable. Now, the surveillance platforms can sometimes destroy a target. But, any MW capable adversary easily avoids destructive drone strikes through infrequent use of personnel carriers, and keeping its infantrymen well dispersed. Drone missiles are far too expensive to waste on a single squad. And, the last minute dispersion of each squad would sufficiently protect their members. In the daytime, a drone in Ukraine would have no more luck eliminating a entire squad than an attack helicopter did in Vietnam. The day when clouds of tiny drones show up—each carrying a smart grenade—are too far into the future to worry about.

In the meantime, ground warfare will remain pretty much the same. The small-unit actions that allowed the Unites States Marine Corps (USMC) to win WWII in the Pacific could also save the Ukrainians. But, the biggest threat will still be long-range MG fire. Today's hand-held-munitions make armored personnel carriers a bad way to try to escape it. Besides the latest shoulder-fired antitank weaponry, the Russians have a SCHMEL thermobaric rocket with the blast equivalent of an 155mm artillery shell. To the new U.S. Javelin armor destroyer, the Ukrainians must only add some shoulder-fired anti-aircraft missiles. Then, hundreds of semi-autonomous assault squads could successfully swarm any Russian outpost, convoy, or forward thrust. For a simulated mortar attack, one or (any U.S. squad) would only need a few concussion grenades. (See Figure 2.1.)

The First Sign of Trouble

Early in June 2023, the Ukrainians attempted a limited push into Russian lines at the south end of the country, as a precursor to their pending offensive. As briefly shown in a U.S. television network video, that push was quickly repulsed. That same news release (or another about the same time) then mentioned that 35 personnel carriers supplied by the U.S. had been destroyed (probably Bradleys).[4] Just what happened to their infantry cargo is not pleasant to contemplate. A Forbes article later that month claimed

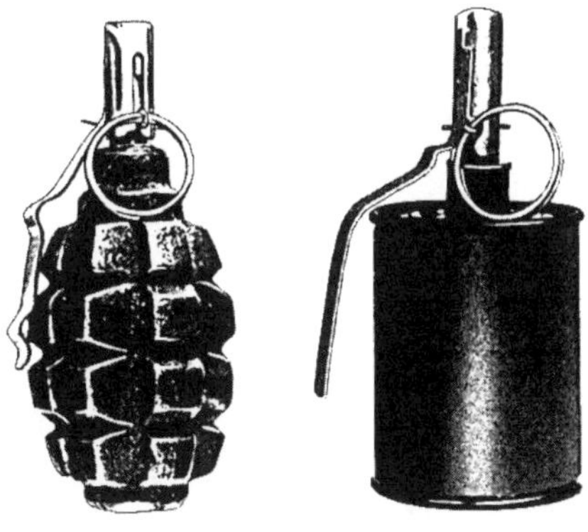

Figure 2.1: PLA Troops Carry Both Frag and Concussion Grenades
(Source: "Handbook on the Chinese Communist Army," DA Pamphlet 30-51 [1960], U.S. Army, figure 72, p. 74.)

8 June as the date of the mishap, and over 20 M-2 Bradley's being lost along with seven Finnish Leopard tanks and a mine clearing vehicle. That article attributed most of the problem to mines.[5]

The Ukrainians had more likely made the same mistake the Russians did on their initial drive into Kiyv. A personnel carrier coming directly at a well-dug-in soldier with the latest bazooka-like weapon is almost as vulnerable as one stuck sideways in a traffic jam. As was demonstrated over and over on Iwo Jima and Okinawa during the latter stages of WWII,[6] armor alone cannot successfully penetrate a good defense network. It has to be accompanied by fully dismounted and highly skilled infantry squads.

The Unfortunate Trade-Off

Various advances in military technology now make individual and small-unit maneuvers (and even closing with the enemy) seem unnecessary. But MW-oriented troops are offensive by nature, and will require good tactical techniques whenever their electronics get shot, wet, or dirty. No new gadget makes the GI harder to detect on the battlefield, or more able to dodge bullets. The lengthy conflicts in Vietnam and Afghanistan have adequately demonstrated that neither U.S. troops nor their allied counterparts should be forced to carry all hard-to-operate and superfluous inventions. They are the by-product of Attrition Warfare—where new equipment seems more important than opportunistic people. MW depends more upon surprise than firepower. That's why America's under-supplied foes were able to win along the ground in both locations.

PART TWO

MANEUVER ELEMENT AGILITY

"TO LEARN WARFARE THROUGH WARFARE—THIS IS OUR CHIEF METHOD."
— MAO TSE-TUNG

(Source: "Mao Tse-tung: An Anthology of His Writings," ed. Anne Fremantle (New York: Mentor, 1962), p. 82.)

3 Single-File Travel

- Which dangers now face an infantry squad on the move?
- What formation best counters those dangers?

Why waste time with fancy formations?

(Source: Courtesy of Sorman Information and Media, from "Soldf: Soldaten i falt," ©2001 by Forsvarsmakten and Wolfgang Bartsch, Stockholm, p. 225.)

Linear Formations Have Long Fascinated Foreign Armies

Across an active battlefield, infantry columns can move faster than other arrays. From the front across level ground, they may look like a single man. (See Figure 3.1.) While highly vulnerable to same-axis bullets, they seldom move along the final protective lines (FPLs)—interlocking bands of MG fire—of an old-fashioned "sentry string" defense. Only when attempting a path of least resistance do they normally get into trouble—like when squad strongpoints of the WWII Japanese aimed their MGs down a draw or finger. On-line formations are far more dangerous in the assault. Flanking

German gun pits mowed down whole waves of French and British troops during the first part of WWI, and up to 33,000 U.S. Doughboys near its end.[1]

Overall History of the Infantry Column

Throughout the U.S. Civil War, Union troops traveled three abreast. Whenever enemy contact was imminent, the exterior files veered off at right angles, and the center one fanned out.

Late in WWI, at the point of every German offense was a Stormtrooper squad column. Never aimed directly at a defense line MG, that squad spear breached its barbed wire with a bangalore torpedo amidst shifting artillery shells. Once inside Allied lines, it turned sideways to attack that bunkered gun with concussion grenades and a satchel charge. Meanwhile, the Western Powers experimented with "flying columns" formed after a foe's front lines had already been breached to more quickly access his rear areas.

Then, Japanese assaults on Guadalcanal's Henderson Field at the start of WWII bore a striking resemblance to the German Stormtrooper method of WWI. They entailed: (1) sneaking up to the wire; (2) bangaloring it during an artillery barrage; and then (3) negotiating the gap with bayonets and grenades only.[2] Most Japanese platoons closed with the American perimeter through a row of squad columns.[3] In that way, any breakthrough could be exploited by sister squads.

During the Korean War, a row of extended Stormtrooper-like squads were to mimic Chinese human waves at the Chosin Reservoir.[3] Upon reaching U.S. lines, those squads took turns grinding away at them. Some may have carried a bazooka-like weapon for MG bunkers. Their approach march had been done in a gradually winnowing queue.

> Unseen and unheard [after sunset on 27 November], the endless columns of quilted green [actually white] wound through valleys and over mountain trails leading toward the southern tips of Northwest and North Ridges. These were the assault battalions of the 79th and 89th CCF [Chinese Communist Forces] Divisions.[4]
>
> — *U.S. Marine Operations in Korea,* by Montross
> History Branch, HQMC

Figure 3.1: Personnel Array with the Least Movement Signature
(Source: FM 7-8 [1984], p. 3-52.)

[CCF] regiments, attacking in columns of battalions, deployed in[to] columns of companies [and then possibly columns of platoons].[5]
— *Chosin,* by Hammel

At 2125 the mortar eruptions began to walk toward the Marine rear. Whistles screeched, enemy machineguns fell silent, and the first Chinese assault waves [a single row of extended squad columns] hurled themselves against the juncture of Companies E and F. The enemy attacked on an extremely narrow front [one squad at a time] in order to maintain [random] control. His troops advanced in column within grenade range.[6]

– *U.S. Marine Operations in Korea,* by Montross
History Branch, HQMC

In Vietnam, the NVA would attack a U.S. perimeter through side-by-side squad columns, each with its own rocket-propelled grenade (RPG) man in the lead.[7] The parent-unit advance to the staging area had also been accomplished through piecemeal files. Most often observed at dawn or dusk, such a file passed by a 27th Marines Headquarters listening post [LP] around midnight in the Spring of 1968.[8] Through such sightings, one concludes that most approach marches of Chinese Mobile Warfare would likely occur during periods of deep shadow.

Other Column Attributes

Back-to-back chains prevent unit members from getting separated at night or in deep vegetation. So arranged U.S. troops usually keep an eye on the man to their front. Carlson's WWII Raiders had an eight-foot toggle rope with which to guarantee the link. But, Rogers' Rangers monitored the next boat or man to their rear during the French and Indian Wars.[9]

Stealth and less human sign are easier to maintain while in single file. Successive members remove any noisy obstruction or step into each other's footprints, while rear guards brush out all hints of passage.

Columns can also add tactical opportunity to any maneuver. Alternating which direction ambush patrol members face while stationary produces 360° protection. In Vietnam, the foe often used a parallel-line defense. If about to be overrun, the forward line could fall back on its partner. And, by remaining in column, assault troops can fire in any outward direction without risking fratricide. Then, by becoming flying columns, they can more quickly consolidate an objective.

Movement String Options

When all column members face to the right, left, or rear, they can march as one in any direction, and thereby change location through one smooth effort. (See Figures 3.2 and 3.3.) Most U.S. small-unit maneuvers are not as easy to accomplish. GIs regularly practice every manner of security pattern, to include unit wedges, "V's", and circles. That much precision is seldom necessary in combat. And, it sacrifices any momentum developed.

Far more useful is a column or file that easily turns into a line or row. During the crossing of rough ground, changing from line to column and back again was stressed by the Japanese in WWII. Also practiced was how to the change the direction of a squad on line.[10] Without constant practice, groups of soldiers can't execute the simplest maneuver without becoming disoriented.

Among the column-shaped attacks is the Indian-file night attack. A human chain sneaks up to enemy lines and then crawls through at an angle. At some point, everyone stands up and assaults toward the center of the objective in echelon formation. Through "recon pull," the assault force has eliminated the standard night attack's telegraphing of intentions. There are no more day and night reconnoiters, enemy line sentinels, guided approaches of separate increments, or further alignments before the assault.

Eastern Army Units Only Use Forward Scouts

Instead of a security element at every quadrant and leader doing the land navigation, Eastern maneuver elements rely almost entirely upon point scouts. Those scouts choose the final route and watch for trouble. (See Figure 3.4.) To keep from getting spotted, they take advantage of all shadows, vegetation, and folds in the ground. They know that U.S. defenders pull in all outposts at the first sign of trouble, so as not to mask their own fires.

To more easily sense an ambush, each scout moves well ahead of the main body. Even highly disciplined soldiers make telltale sounds in an ambush site, and the Eastern scout has *ninjutsu* training on how to better hear them.

If an ambush is suspected, he directs the main body to one side. From there, part can sweep through the ambush formation from a flank. (See Figure 3.5.) That assault need not be done standing up. Its fire alone will discourage the most resolute ambusher. Then, the friendly column resumes direction, eventually assuming the original

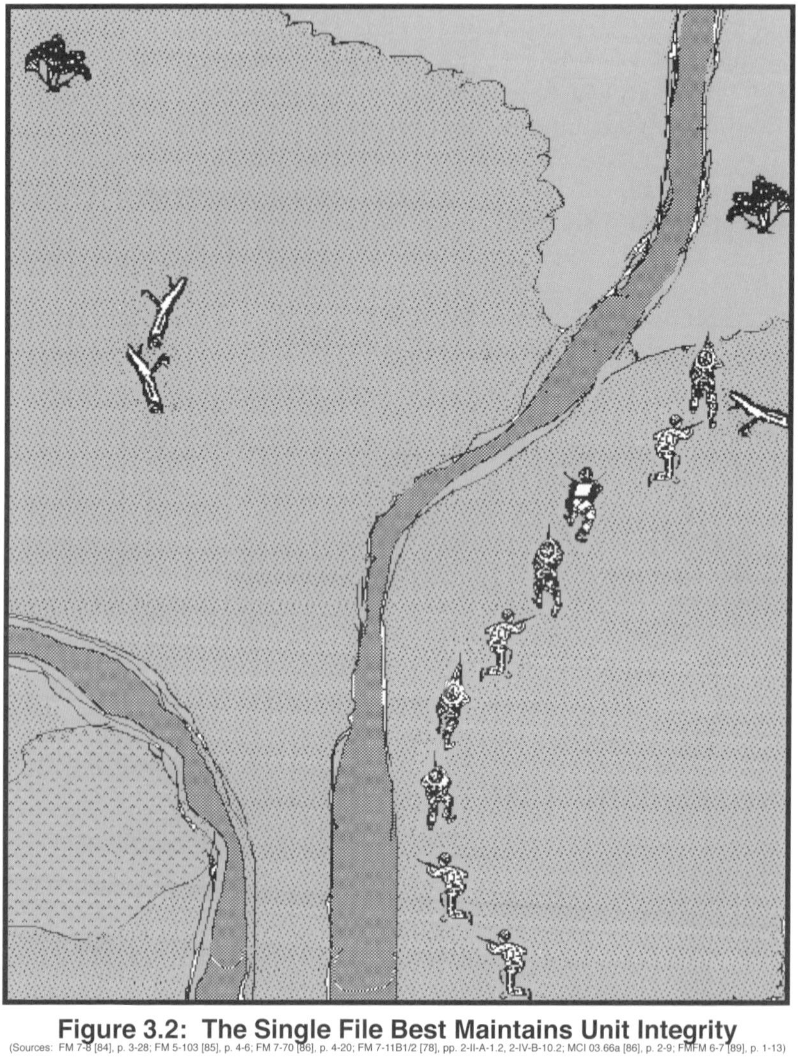

Figure 3.2: The Single File Best Maintains Unit Integrity
(Sources: FM 7-8 [84], p. 3-28; FM 5-103 [85], p. 4-6; FM 7-70 [86], p. 4-20; FM 7-11B1/2 [78], pp. 2-II-A-1.2, 2-IV-B-10.2; MCI 03.66a [86], p. 2-9; FMFM 6-7 [89], p. 1-13)

Figure 3.3: Facing Left Creates Proper Chance Contact Alignment

(Sources: FM 7-8 [84], p. 3-28; FM 5-103 [85], p. 4-6; FM 7-70 [86], p. 4-20; FM 7-11B1/2 [78], pp. 2-II-A-1.2, 2-IV-B-10.2; MCI 03.66a [86], p. 2-9; FMFM 6-7 [89], p. 1-13)

Figure 3.4: Point Spots Enemy Ambush
(Source: FM 7-8 [1984], p. 5-22.)

route. Land navigation by terrain association will be more useful at this stage of the game than Global Positioning System (GPS) monitoring.

Rest Stops

A good defense can be instantly established after traveling single file. (See Figures 3.6 and 3.7.) Such a formation permits noiseless sleeping. No flank outpostings are necessary if the ground is well-enough selected. A nearby berm or log would provide adequate protection from enemy fire.

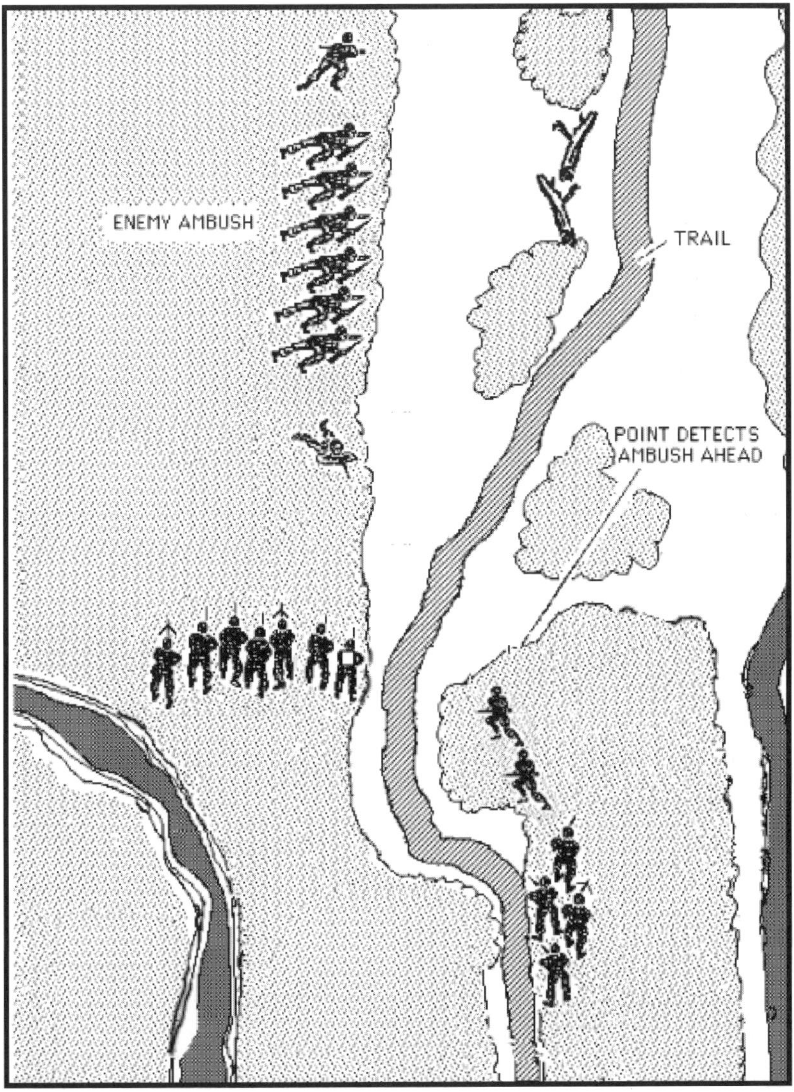

Figure 3.5: Column Turns Left to Attack Ambushers from a Flank
(Sources: FM 7-70 [1986], p. 4-20; FM 5-103 [1985], pp. 4-6, 4-38; FM 7-11B3 [1976], p. 2-VII-C-4.4.)

Other Common-Sense Requirements

One's degree of wartime safety is in the eyes of the beholder. It's now widely admitted that Attrition Warfare results in more friendly casualties than any 3GW variant.[11] The reason probably lies in how enemy MGs are handled.

Most Japanese gun bunkers on Okinawa during WWII were so well hidden in broken terrain that U.S. observers had to be within 20 yards of one to accurately summon artillery shells. Even when those shells could be walked in from the back, they posed a friendly risk and seldom penetrated the target.

> Even the . . . 155mm shells . . . required ten [10] to 12 hits, all in the same place, to inflict major damages on many of the imposing Japanese fortifications (VAC ArtpRpt 29).[12]
>
> — *Iwo Jima: Amphibious Epic,* by Bartley
> Historical Branch, HQMC

Figure 3.6: A Hasty Defense Is No Problem
(Sources: FM 5-103 [1985], p. 5-2; FM 21-75 [1967], p. 67)

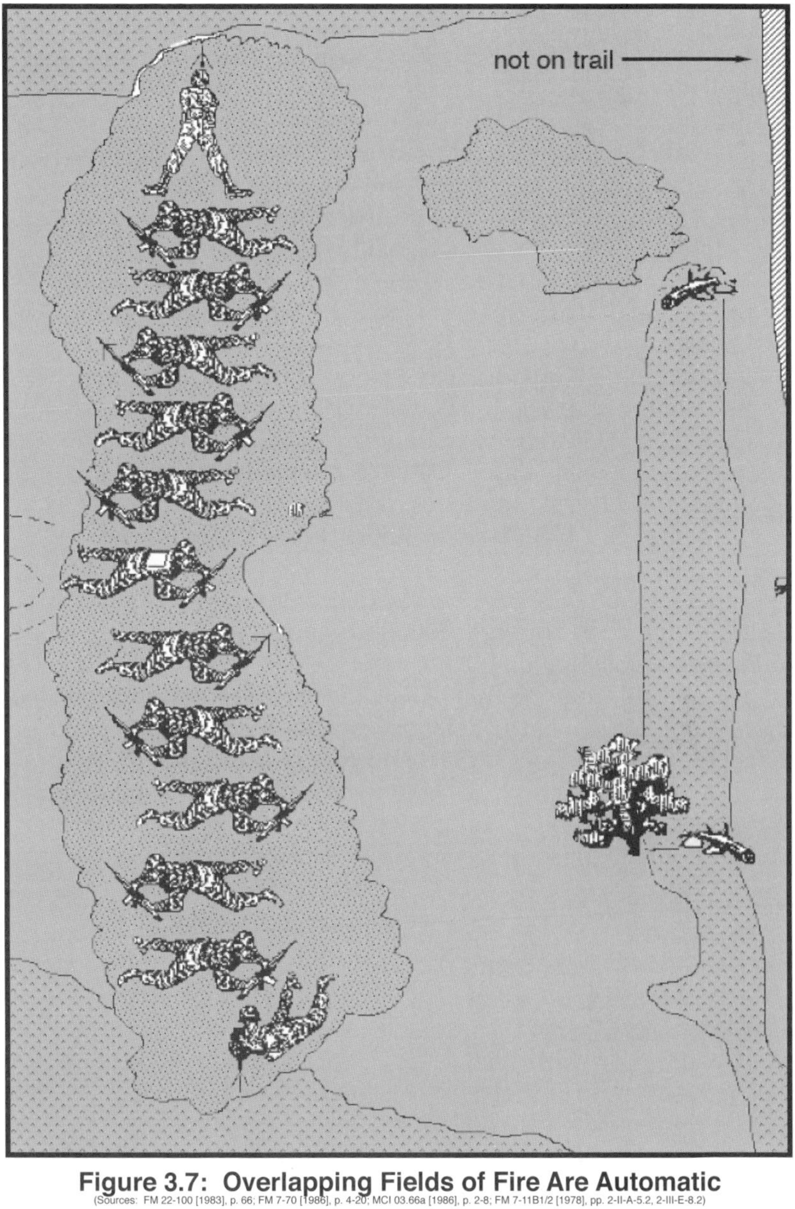

Figure 3.7: Overlapping Fields of Fire Are Automatic
(Sources: FM 22-100 [1983], p. 66; FM 7-70 [1986], p. 4-20; MCI 03.66a [1986], p. 2-8; FM 7-11B1/2 [1978], pp. 2-II-A-5.2, 2-III-E-8.2)

America had planned to take Okinawa through 2GW—with ineffectual bombardment as its weakest link. Such a paradox was hard to quantify after an unopposed landing. But, it became a mathematical certainty when three fully supported U.S. divisions tried to blow their way through the first enemy barrier—the Machinato Line.

> The assault platoons advanced [on 19 April 1945], hopeful that the great mass of metal and explosive had destroyed the enemy or had left him so stunned that he would be helpless. They were soon disillusioned; for the Japanese, deep in their caves, had scarcely been touched, and at the right moment they manned their battle stations (XXIV Corps Actn Rpt, 37). Brig.Gen. Josef R. Sheetz, Commanding, General, XXIV Corps Artillery, later said he doubted that . . . [more than] 190 Japanese, or 1 for every 100 shells, had been killed by the morning artillery preparation (Appleman, *XXIV Corps History,* 179).[13]
>
> — "Okinawa: The Last Battle," by Appleman et al
> *U.S. Army in WWII Series*

Japanese bunkers were even hard to destroy when hand-tossed explosives made it through their gun ports. Many had grenade sumps, blast doors, and fallback rooms.

So, with all that risk from long-range MG and artillery fire, too much worry about nearby ambushes would hinder U.S. maneuver elements. Shouldn't they try just as hard to avoid standoff barrages? A unit with all-around security can't move quickly enough to do so. In the modern age of drone-directed precision fire, only separately traveling squad-sized components can last long enough to conduct a joint attack.

4 What Requirement for Flankers?

- Why don't big Eastern armies use flank security guards?
- What are the guidelines for their use in the U.S. military?

What type of terrain is most dangerous without side sentinels?

(Source: FM 7-8 [1984], p. B-2)

The Generally Accepted Rule

No U.S. grunt questions the need for flankers. On patrol, he has always used them, and his manuals stressed them. Since WWII, American recruits have been punished for not instantly complying with any order. But, that was when all tactical doctrine was of 2GW origin. Now, it has to do with 3GW (avoiding the foe's forward forces to get at his rearward centers of gravity).

And, not many American infantrymen are happy with the plodding and predicable way his unit fights. Why don't the big Eastern armies operate that way? There is no mention in their manuals of maneuver element flankers. For an increase in speed, those elements often travel "several columns abreast" with only frontal security (point men). "[Russian] scouts will move through small woods and brush, while the bulk of the squad goes around."[1] So, this over-emphasis on side security personnel in the American military demands further explanation.

Why Flankers Were Invented

Side security men theoretically protect the parent unit from surprise ambush. But, the only terrain demanding all-around movement protection is a big open area. If pairs of flankers don't walk both sides of that clearing as their parent unit crosses its center, the entire unit could be annihilated. And, the same thing might happen on a long straight trail, if a point team doesn't first go to the next bend. That's where the definition of "danger area" comes from.

But, not every gully or draw must be as carefully traveled. With the advent of thermal imaging, moving more quickly through the landscape with nothing more than a point is fully justified. Just staying in "hull defilade" can keep the unit from being spotted early from ground level. And, a good point man will hear the twig snap, rifle butt bump, or AK47 come off safe of a waiting ambusher. As long as his unit only uses trails as reference points, its risk of ambush will be minimal.

During WWI, the German Stormtroopers approached Allied lines through the shallow communication trenches that crisscrossed every battlefield. A 1997 Russian manual describes how properly to traverse uneven ground under fire—run bent over through depressions and crawl (or roll) over rises.[2]

Nor does thick trail-side brush require side security. In early 1967, one platoon was the point element of 1st Battalion, 4th Marines on the trail between Gio Linh and Con Thien near Vietnam's very dangerous Demilitarized Zone (DMZ). When a big bramble bush blocked any passage to the north of the trail, the two right-side flankers asked to come in. If the platoon leader hadn't let them, both would be dead. Unbeknownst to anyone at the time, an interior leg of the Ho Chi Minh Trail intersected the main east-west conduit at that point,[3] and a southbound NVA unit had been waiting to cross. The Marines couldn't be accurately shot because their narrow pathway was slightly sunken and obscured by hedgerow. When the lead NVA element did finally open up with its light MG, it was unable to accurately target any Marine.

The Modern-Day Shift in Emphasis

U.S. ground assaults against enemy objectives have become riskier as thermal imaging makes their approach marches easier to spot from the air. But, that doesn't mean that Americans have to fight more defensively, or solely depend upon their standoff firepower. As the Battle of Okinawa has conclusively shown, heavy bombs and shells won't dislodge a well-dug-in defender. He still has to be engaged along the ground. The way to do so safely is to travel and attack in piecemeal fashion. Thermal blankets now help grunts to escape detection from above. And, historical events reveal other ways to disappear.

The best known example is the Chinese Army's secret move into the Chosin Reservoir in late 1950, where no fewer than 300,000 Chinese volunteers secretly journeyed hundreds of miles through territory under almost constant U.S. aerial surveillance. They traveled mostly at night or under the cover of forest-fire smoke, and then hid during the day. Whole platoons would cram into a single-family dwelling or hide under whatever nature provided in a draw or gully. When forced to move during the daytime hours, they carried straw mats to cover themselves. Then, by closing ranks and lying down in formation, they could simulate a freshly cut farmer's field. While dressed in white, refugee mufti, or Republic of Korea (ROK) uniforms, some moved openly along the roads.[4]

> In rear and assembly areas, CCF troops dug two-man or squad-sized foxholes on reverse slopes. The soldiers carefully distributed the evacuated soil, covering it with branches and straw and replanting the turf. These foxholes were cunningly selected and camouflaged to blend in with the [surrounding] terrain; they were virtually undetectable except from close range (U.S. Army, *Engineer Intelligence Notes,* 1951, 8:4). The Chinese also selected bivouac sites in or near burned-out villages, often taking up residence among the rubble. The CCF avoided high ground, which characteristically attracted air strikes (U.S. Army, *Enemy*

> *Tactics, Techniques and Doctrine,* 1951, 5). In addition, the Chinese used natural materials almost exclusively for camouflage and concealment. . . . [United Nations] U.N. patrols sometimes approached within killing range of the CCF, but the Chinese often did not fire, choosing to remain hidden.[5]
>
> — U.S. Army, *Leavenworth Research Survey No. 6*

A Few Longtime U.S. Rules Have Remained Useful

For a unit in transit, the traditional definition of a danger area was one where the entire unit could be destroyed. To remove that risk from any piece of terrain, the unit's commander had only to send a few upright people across it at a time, or have everyone crawl over it. This exception equally applied to a heavy vegetated chokepoint or open clearing. That's how a maneuver element or patrol could negotiate either obstruction without first checking for

Figure 4.1: Any Light Background Creates an Artificial Skyline
(Source: MCRP 3-02H [1999], fig. I-3.)

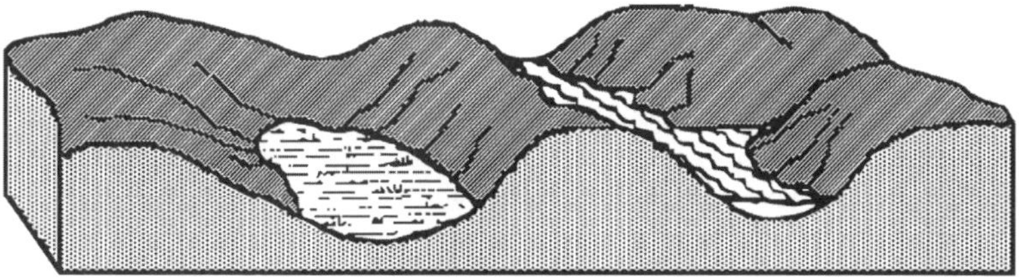

Figure 4.2: Ground Relief Can Be Helpful or Problematic
(Source: FM 21-76 [1970], p. 49.)

enemy soldiers. It is also how a unit only controlled by its forward scout could approach opposition lines with a minimal movement signature.

The pre-WWII Japanese enjoyed written instructions on how to conduct a night movement: (1) avoid entering woods or villages when possible; (2) keep a low posture in open terrain; (3) walk in the shadow along roads, treelines, and streets; (4) don't all start or stop at once on gravelly ground or a descending slope; and (5) lie down and listen at any suspicious indication.[6] They knew that slow movement across a bushy field in low light was hard for the human eye to detect. They were also careful to stay off any skyline. (See Figure 4.1.)

The Chinese approaching the Chosin Reservoir exploited all types of terrain, big or small. (See Figure 4.2.) They followed linear terrain features to get where they were going—draws, fingers, power lines, firebreaks, and ridgelines. Then, they used watershed gullies to move beneath the Marines' MG FPLs.

> Chinese infantrymen [in Korea] also knew how to use every fold in the terrain for cover and concealment. They seldom got disoriented in the darkness and showed an unerring appreciation of ground in the choice of their routes.[7]
>
> — U.S. Army, *Leavenworth Research Survey No. 6*

Road and Trail Danger

When man-made features are only used as reference points, they don't pose much of a threat. One can keep track of them from a distance. And, any straight-line travel across a contested area has very little chance of enemy contact, as long all lines of drift are avoided (like a flat streambed through precipitous terrain). The only real drawbacks of a dead-reckoning journey are the time-consuming obstacles that might have otherwise been avoided. Yet, this is one occasion when a GPS device might help to determine one's final location.

But land navigation by terrain association also works as long certain landmarks aren't used as checkpoints: (1) trail or road junctions; (2) tall objects with a good view; (3) major draw or ridge intersections; (4) the very end of a big draw or finger; and (5) a stream ford or bridge. Better choices would be: (1) any place at which the desired route hits (but does not cross) a road or stream; (2) where two different linear terrain features intersect (like a trail and draw); and (3) the near side of any map-indicated break in the vegetation (like a power line corridor).

The wary traveler moves just above the military crest of a finger or ridgeline (and never along the topographical crest). That way, he won't be silhouetted from below, yet still enjoy good observation and fields of fire over anyone there.

Even at night, friendly patrols must stay well clear of barren ground, standing water, and light backdrops. They should gravitate toward jumbled ground, deep shadow, and dark rearward settings. That's because a few objects in an otherwise open area tend to draw unwanted attention; water reflects movement; and all light backgrounds act as an artificial horizon after dark. From starlight alone, asphalt roads appear almost white at night. "Want-to-be" shadow walkers sometimes forget this. Yet, any long artificial horizon can also act as an asset. The unit wise enough to move parallel to—but just barely within sight of—a road can often spot enemy moving along its very edge without being noticed itself.

Speed

To call artillery fire on a moving foe unit in in Vietnam, U.S. troops first had to get headquarters' permission. This took time.

Periodically, U.S. units risked such a mission without clearance. One time south of Da Nang, the commanders of collocated infantry and artillery outfits shot at a string of bushes they saw running 4,000 yards out at dusk through their "BC Scope" (long-range Navy binoculars). Even after a perfect spotting round and immediate "fire for effect," they could only pepper the woods just entered by their target with 105mm rounds.

Where enemy supporting arms threaten, speed acts as a type of security. As would the North Vietnamese,[8] the Chinese in Korea often ran in column during an approach march. Their scouts provided all the directional guidance necessary.[9]

Speed and "Recon Pull" Are Part of the New U.S. Doctrine

Letting one's scouts choose the precise route is the same as "reconnaissance pull" in MW. Thus becomes clear why Americans can't fully practice their new doctrine. They still have too many old habits. In effect, their manuals and schools still teach Attrition Warfare methods. By focusing inward, they spend more time looking good than being good. While trying to limit the number of firepower accidents, they've lost extra lives, several wars, and their grasp on battlefield integrity. (Look back at Tables 1.1 and 1.2.)

A Whole New Movement Protocol Will Be Necessary

To practice MW, U.S. maneuver elements and patrols will both have to operate differently from before. Neither had tried to move secretly around the battlefield. Maneuver elements followed the same route that reconnaissance patrols had taken to reconnoiter an objective. Security patrols made only a half-hearted attempt to locate opposition forces in their parent unit's tactical area of responsibility (TAOR). And, ambush patrols had failed to limit the sign left moving into an ambush site.

The New Maneuver Element Norm

To bypass enemy strongpoints, U.S. maneuver elements must move more quickly than before and fail to attract the attention of enemy supporting arms. That is most safely done through piecemeal travel. On the outskirts of Seoul, the ideal approach march happened one time by coincidence.

> Company A found itself entering the main street of Yongdungpo totally unopposed. "It was eerie," said Barrow. "We simply slithered into town undetected."
>
> The 87th [North Korean Peoples' Army] NKPA Regiment, desperately attempting to patch together a defense in depth, had accidentally left this critical approach unguarded, and Barrow took full advantage of the opening. His 200-man company flowed rapidly into the heart of the city, sweeping up surprised bands of the enemy in the process. Before dark they had cut the city in two. Barrow selected a sausage-shaped dike, 30 feet high and 150 yards long, as the place to make a stand for the night. "We immediately recognized that we had a valuable piece of real estate," he said. From the dike his Marines could interdict the intersection of the highways from Inchon and Kimpo.
>
> Through this intersection . . . marched a large formation of unsuspecting NKPA infantry, singing political songs as they hurried to reinforce Yongdungpo's northwestern defenses. Barrow's interlocking machine guns and 60mm mortars cut down many and scattered the rest.[10]
>
> — "Battle of the Barricades," by Alexander
> History and Museums Division, HQMC, 2000

To breach a strongpoint, the U.S. maneuver element approaches it along a never-used route after some deception. It gets into assault position faster than the foe deems possible. Then, like the Chinese, it mounts a safety conscious holding attack (normally through separate squads grinding away), while infiltrators go after the command post (CP), crew-served weapons, and ammunition dump. It will be using "'ordinary forces' to engage and 'extraordinary forces [infiltrators]' to defeat."[11] Those ordinary forces must be U.S. squads able to run their own assault against a prepared enemy position.

New Security Patrol Protocol

American security patrols must choose checkpoints where the

opposition can be more easily spotted: (1) the ends of straight roads or trails; (2) tall easily climbable trees; (3) edges of large clearings; (4) established stream fords; and (4) thinly forested flat land. They must also start looking for enemy footprints, and then get permission to follow the trail.

Any U.S. security patrol that comes back to is base camp without making contact in an enemy infested area has failed at its mission.

New Ambush Patrol Methods

Future American ambush patrols will be like phantoms. They won't be seen leaving the friendly perimeter, anywhere on the move, or while entering an ambush site. Nor will the quarry ever know it was ambushed. It will think the source of all explosions to be land mines or mortar fire. No longer will U.S. forces just try to intimidate the enemy. They will instead suspend all opposition travel across occupied territory.

5 Dodging Foe Outposts and Roving Sentries

- By what route can an assault unit best dodge foe outposts?
- How much surprise can be lost for an attack to succeed?

Roving enemy sentries are harder to spot than wild animals.

(Source: "A Concise History of the Unites States Marine Corps 1775-1969," by Capt. William D. Parker, Hist. Div., HQMC, 1970, sketch by Capt. Donald L. Dickson, p. 62)

Avoiding Lines of Drift *en Route* to the Target

Land navigation by "terrain association" follows a series of linear ground features to the final destination. But, any gully or finger near an attack objective will constitute a likely avenue of approach, and likely have an Eastern MG aimed down it. By dead reckoning along any compass heading through a heavily vegetated landscape, one has little chance of making enemy contact. He can also avoid other lines of drift (like a flat streambed through precipitous terrain). (See Figure 5.1.) But, if his present location is known to the enemy, a two-leg zigzag would be less well guarded, as long as the center checkpoint is obscure or only GPS identifiable. Such a checkpoint might be: (1) where the route hits (but does not cross) a road or stream; (2) the point at which a trail and draw intersect; or (3) the near side of a map-indicated break in the vegetation (like a power line corridor). During the final approach, no enemy will adequately outpost swampy ground.

No forward reconnaissance can leave much evidence of passage near the intended point of attack. Eastern defenders regularly look for human sign around their defensive positions and can backtrack it.[1] The attack unit will enjoy more success if not ambushed on its way down to the objective.

Figure 5.1: Dead-Reckoning More Useful Than Once Thought
(Source: FM 90-6 [1980], pp. 1-4, 1-5.)

The Enemy's Use of Roving Sentries

While on defense against a world-class opponent, the American unit must be as devious possible. The WWII Germans often employed two-man "reconnaissance and visiting" patrols within each row of defensive sentry posts. Their job at night and in broken terrain was to observe the intervening areas not occupied by posts, and perform liaison.[2]

The Japanese of that era did likewise. In fact, the 1913 Japanese night-fighting manual had listed "not staying stationary" as one of the conditions for a good LP.[3] As the night sentinel often had to shift location, he was further trained in night reconnaissance.[4] At one point during the Nomonhan Incident of 1939, Japanese defensive scouts were able to shadow Russian infiltrators.[5] While moving, night sentinels were taught to: (1) not make noise or cast shadows; (2) not move quickly; (3) not go in the wrong direction; (4) not approach another sentinel unannounced; and (5) not compromise a passing patrol.[6] In fact, the patrols and entire sentinel line were required to share with one another everything they had learned about the enemy.[7] That is a perfect example of how much more can be accomplished by decentralizing control. Western commanders would never dream of operating this way. The Russian infiltrators must have been equally skilled in the Russo-Japanese War of 1904-05, because the subsequent Japanese manual warned of Russian interference during the relief of a sentinel.[8]

During WWII, the daytime Russian sentry post was normally manned by two stationary pickets. But in Afghanistan, the Russians operated differently. "[Their] security outposts functioned around the clock. During the day, one man per squad or tank was on watch while a two-man patrol worked the area."[9]

In Manila near the end of WWII and again in Seoul during the Korean War, U.S. troops encountered lone enemy sentinels inside otherwise empty buildings.[10] Those sentinels would have regularly moved around their respective structure, and those in both places likely doubled as forward observers for supporting arms.

Maneuver Element Signature

Now, every unit on the move has a sight, sound, heat, and electronic signature to worry about. During the straight-line approach to an attack objective, the first would require crawling across any open area. The second could be handled through stealth. The third would't be an issue in the daytime, but might be possible through thermal blankets at night. And, the fourth may take leaving all cell phones and certain other devices behind. Otherwise, the journey may be noticed by drones. The most promising alternative would be an old NVA trick from Vietnam—make the trip piecemeal along several different routes and then reassemble briefly for the attack.

6 The Elsewhere Located Base of Fire

- How much can a detached base of fire help one's assault?
- What's the purpose of its stream of bullets?

How dangerous to friendlies is a U.S. base of fire (BOF)?

(Source: U.S. Army Center of Military History, artphoto archives, illustration designator "avop08-01.2.jpg," An Army of One, by Henrietta Snowden)

Fire Superiority Always the Goal with Attrition Warfare

Western units are supposed to set up an elsewhere located BOF for every assault. The intent is good—to deceive the foe and limit his ability to respond. But, the result is a ruse that comes as no surprise to the quarry. While a mortar barrage can be helpful (as for the concussion-grenade throwing German Stormtroopers), MG bullets are frequently a waste of time. Normally from a direction other than the assault route, those tiny projectiles don't easily enter sideways facing gun ports. Only an overhead BOF will do crawling assaulters any good, and it is difficult to coordinate. That's why most Eastern armies collocate their MG's with the assault element.[1] Then, at least, the opposition's most destructive MG can be accurately targeted.

Other Armies Make MGs More Available on the Move

While on a patrol or approach march, every German squad's MG team walked near the front of a single-file formation.[2] Should this formation make chance contact, that team formed a skirmish line with half of the maneuver section on either side. Those three elements then alternately moved forward in response to the squad leader's whistle.[3] Or, any number of riflemen took turns advancing at his command.[4] Rushing individuals would roll to one side after dropping to the ground to disguise their final location. Smoke and concussion grenades were subsequently tossed during the final assault.

During a Russian patrol or approach march, the MG personnel also walked near the front of a column.[5] During any hasty attack, separate parts of the maneuver element advanced on either side of a BOF element. At the squad leader's direction, all three moved forward by "team" or "individual" bounds.[6] Rushing soldiers would flop down near hollows, shell holes, or other cover.[7] They dropped like stones and then crawled a short way.[8]

Something similar was to form the basis of the U.S. Ranger hasty attack against a German gun pit behind Omaha Beach. One GI went to left, one right, and a third toward the center to draw the occupant's attention.[9] The U.S. infantry squad has never an organic light MG.

Bunkers Can Be Nearly Impervious to Enemy Fire

As Chapter 1 discussed, bunkers can be so well built that direct hits from heavy ordnance won't penetrate their sides and explosives through their gun ports won't silence their MGs. So, why would anyone think that a distant MG would stop defenders from sticking AK47's out of holes or around corners to mow down approaching GIs? Those in charge need to let a combat Private edit their next infantry manual. The whole idea that a BOF easily achieves fire superiority is an Attrition Warfare myth. That's why short-range infiltration is now the standard offensive maneuver for Unconventional Warfare (UW). There's no longer a noisy BOF to warn the quarry of an impending assault.

A Fully Detached BOF Has Its Limitations

While lifting the siege of Thuong Duc during Operation Maui Peak in the autumn of 1968, 2nd Battalion, 5th Marines was sent as a diversion to the mouth of a mountain valley south of Da Nang. While trying to take a strategic peak at the foot of Charlie Ridge, Golf company ran into unexpected problems with regard to an elsewhere located BOF.[10] After climbing a nearby spire by accident, the unit left two MG's behind to cover its assault should the proper destination be occupied. It was. And, as a single platoon began its assault up a brush-covered slope that was fully visible to the MGs, the company commander behind the on-line formation took overhead MG fire. He dropped to one knee and radioed for the BOF to stop shooting. Then, while the rest of his unit was safely below the military crest of the hill, every bush to his rear was cut in half by another burst. Both had come from an enemy gun. But, the lesson soon became clear. Even with an unobstructed view of the attack route, that BOF would have had trouble shooting just ahead of the assault line. And shutting all supporting fires off early left fully exposed troops largely to the mercy of the enemy. Luckily, only two Marines would be killed in two back-to-back (and ultimately successful) attempts at this particular hilltop.

7 Are Stand-Up Assaults Absolutely Necessary?

- When can upright soldiers safely rush an attack objective?
- How often do undetected GIs get that close to defenders?

Should an assault ever cost a squad a third of its strength?

(Sources: FM 23-30 (1988), p. 5-3; FM 7-8 (1984), p. 3-28)

Where the Bullets Can Be as Thick as Bees

By the time WWII rolled around, almost every battlefield event had to do with MGs. One side had increased its pre-assault bombardment to try to silence them. The other had allowed more prone movement during approach marches and final assaults to limit their effect. The German infantry manuals actually encouraged crawling during an attack.[1] And, for all other Eastern armies, squirming along the ground had become noble assault procedure. While the Germans and Russians secretly crossed lightly held enemy sectors using long-range infiltration, the Japanese and Chinese snuck through heavily guarded ones with the short-range variety. As WWII drew to an end, the Japanese stopped running *banzai* charges altogether.[2]

The General Rule for "Bum Rushing" an Opponent

Finnish manuals still contain the prerequisites for a successful stand-up assault. Participants must be within grenade range of their quarry and toss concussion grenades before moving. This logical a progression may have come from the German Stormtroopers of WWI.

But, most Western armies don't focus that much on their opposition. They don't feel it necessary to surprise every defender. By the time a ground assault is conducted, they assume he has already been disabled by bombardment. That's why U.S. manuals don't specify how close upright attackers should be to their quarry or their assault speed. The system has tried so hard to achieve fire superiority, that assaulting GIs will hesitate after realigning inside a foe's barbed wire. Nor will they have concussion grenades to keep the defenders' heads down. That's a crying shame, because approaching a fully alert and partially hidden enemy machinegunner—while fully exposed oneself—is not a good idea.

Total Surprise Is a Must

Young thugs wishing to jump an opposing street gang must be almost within arm's reach of their target. Without a diversion, no one any farther away will shock an opponent. On the battlefield, concussion grenades can serve that purpose quite well, but bases of small-arms fire are hard to turn off at the right moment. As with a rolling artillery barrage, the distance between suppressive missiles and assaulting troops often becomes too great to keep the GIs healthy.

So, the best prepared of the world's armies—those from bottom-up cultures—have developed other ways for infantrymen to assault than was used on Operation Maui Peak in the last chapter. Slightly more complicated than the U.S. standard, those ways also better mesh with frontline realty. One of the most interesting was the probing assault of the Chinese at the Chosin Reservoir in late

1950. While designed as a holding attack to draw the Americans' attention away from rearward infiltration, this cleverly designed "diversion" was powerful enough to continue forward and seize several objectives by itself.

During the "heavily disrupted" mass attacks of 27 November 1950, Chinese regiments assaulting Hills 1240 and 1282 still managed to take both objectives.[3] Other regiments not only displaced How/7 from Hill 1403 on the southeastern end of Northwest Ridge, but also Charlie/7 from Hill 1419 on the road to Hagaru,[4] How/5 from Hill 403 at the head of the Yudam-ni valley,[5] and elements of Item/5 from Hill 1384 at the southwestern end of North Ridge.[6] How many more Marine companies were temporarily split as Communist forces passed through them may never be known.

Non-Suicidal Grinding at the Chosin Reservoir

Japanese infantry regulations of 1913 had required attacking at night with "a line of company columns."[7] Most of each company would have hung back to see how its forward element fared. Then, just as the Japanese had assailed Guadalcanal's Henderson Field with a row of squad columns in 1942,[8] so would the Chinese first attack Marine positions at Korea's Chosen Reservoir with a brace of squads eight years later.[9] Each Red battalion column had split into company columns, and then platoon columns. Whether each platoon further fanned out is unlikely. More probably, it remained a string of back-to-back squads. Each Chinese squad contained three fireteams with three men apiece,[10] for a total of 10 men.[11] There exists a Communist photograph from the rear of a single file of over 25 soldiers who has just made contact with an American hilltop position.[12]

At the very front of every CCF battalion had been at least one infantry squad specially trained at probing U.S. lines.[13] So, each platoon may have had its own counterpart. That specialty squad had crawled to within grenade range of Yank foxholes.[14] Then, members of its lead fireteam came abreast to keep from getting shot by the same bullet.[15] If the attack had been prematurely discovered, similar spreading may have occurred at every echelon. Once under illumination, Leathernecks saw four rows (15 yards apart) of side-by-side Asians.[16] But, those had not been four consecutive on-line companies. Back-to-back elements of the same platoon had more probably aligned with those from other platoons. That way, any Marine perimeter breach could be exploited by members of the same unit.[17]

All frontal Red Army attacks were then of the more "safety conscious" probing variety.[18] After each breach attempt would come a partial withdrawal covered by the next squad back. That way subsequent assault elements could avoid previous mistakes. (According to Chairman Mao, every size of Communist element learns through fighting.)

The Chosin Marines had very little barbed wire, so it may have been clipped by the same scout who led each East Asian platoon into position.[19] While most Red soldiers had mortar-simulating concussion grenades and bamboo bangalores,[20] that lead scout was a trained infiltrator who would have tried to enter Marine lines before any noisy initiative.[21] His initial target would have been a crew-served weapon. But, if that entry attempt had failed, he may have moved to some other location. According to a U.S. Army study of the Korean War, the Chinese routinely held a defender's attention with an overt frontal attack while sending covert elements around both flanks.[22] Wherever possible during WWII, attacking Nipponese units also attempted a short-range infiltration from a different direction.[23]

Some of those rear-side Chinese infiltrators went after U.S. mortar pits and command bunkers.[24] Others assisted frontline comrades. While contesting the human waves, perimeter Marines often reported concussion grenades and "burpgun" fire from the back.[25]

Role of Various Red Weapons in the Grinding

As had happened at Guadalcanal,[26] some Communist fireteam members may have thrown concussion grenades in Korea, while others scampered forward shooting.[27] Or, a few may have wriggled ahead beneath the bursts of rearward burpguns. But, only in Korea would the noise of both weapons seem to alternate, as different elements chewed away at the U.S. perimeter.[28] Within each lane, a series of separate fireteam attacks had apparently occurred with grenades of some type preceding each automatic-weapon-covered advance.

Those Red columns took turns attacking to maintain some semblance of surprise while looking for the weakest Leatherneck sectors (like where units connected).[29] Once such spots were penetrated, the responsible squad ran completely through the U.S. position,[30] and the rest of its platoon (or even company) may have tried to enter through the same gap.

8 Gun Bunker Takedowns

- Have U.S. bunker assaults always been well thought out?
- Which steps have been overly dangerous?

How effective are bunker-busting munitions?

(Source: FM 7-8 [1984], p. B-1.)

Maneuver Prerequisites

To find the best MW way of neutralizing an MG emplacement, GIs must abandon any longtime preconceptions. That's because MW parameters are often the exact opposite of their traditional way of fighting.

The first attempt should be born out of common sense. A deadly stream of bullets will only be avoided by totally surprising the gun crew. That can't be done standing up. Then, an explosive charge must be somehow inserted through a tiny gun port. Often, that port is too well screened by the surrounding micro-terrain for a bazooka-like weapon to arm in time. If the most logical attack ends up looking like established U.S. procedure, it won't be MW. The "blind 'em, burn 'em, and blast 'em" method of early WWII had put far too many people at risk. Not only did the flamethrower man have to run through enemy fire, but so did the demolition carrier. Later in the war, both steps would be accomplished by grenade-throwing infiltrators.

For a tailor-made MW solution, one has only to determine how any big Eastern Army assaulted a gun site near the end of WWII. The Japanese, Chinese, Germans, and Eastern Russians were all from bottom up societies trying to compensate for too little firepower. So, their small unit tactics were more surprise oriented.

The best MW solution will likely involve "recon pull" and "individual initiative." Peleliu's infamous Point gun had been aimed sideways to create a non-apparent wall of lead at the waterline. That gun could only be spotted from a few shoreward directions. To silence it, a beach-front contingent had first to penetrate a squad strongpoint matrix. Then, two of its members had to go out looking for the nearly obscure gun port of an underground bunker. That bunker was so well hidden that shells big enough to hurt it could only be summoned from within their kill radius. The aperture's eventual compromise came through a rifle grenade ricochet.[1]

One of More Interesting Eastern Alternatives

Each Chinese battalion at the Chosin Reservoir in 1950 had a squad specially trained in "probing" enemy lines.[2] Within MW-prone Asia, probing is done to discover a foe's weaknesses, not his strengths. Those special Chinese squads were looking for under-defended U.S. sectors, not MG positions. Just as the German Stormtroopers of WWI, they wanted to enter the American perimeter at a relatively soft place and then move sideways to attack the crew-served weapons from an unprotected flank. Such is the true nature of the MW way of thinking—attacking where the opposition isn't.

Subsequent Role of the Frontline Assaulter

Unlike U.S. grunts, Chinese riflemen were briefed on the tar-

Figure 8.1: RPG Man at the Head of Each NVA Assault Column
(Source: Courtesy of Osprey Publishing Ltd. from "Armies of the Vietnam War 1962-75," Men-at-Arms Series, © 1980 by Osprey, Plate C, No. 2; FM 21-76 [1957], p. 68.)

gets of strategic value inside an enemy perimeter.[3] Once past the first line of foxholes, he was of more value while crossing the rest of the objective on his own (like American Indians were allowed [4]). There were Chinese infiltrators already in that perimeter attacking mortar pits, CPs, ammo dumps, and even MG positions from the back. So, the frontline assaulter grenaded targets of opportunity as he transited the camp. Any more shooting would have risked fratricide.

Fire discipline was also practiced by the Communists in their assault on an American position. In the movie "Platoon," an RPG carrying NVA soldier first crosses U.S. lines between holes (ignoring a defender pretending to be dead in one of them). (See Figure 8.1.) Looking for a subsequent RPG target, that Communist knows his rocket impacts will be hard to distinguish from the ongoing mortar attack. Before long, a satchel-charge carrying comrade runs into a Yank CP tent and blows himself up.[5] While a few Japanese anti-tank techniques of WWII involved suicide, the Communists have shown more interest in each man's survival. This researcher has never seen hard proof of a suicide bomber from a Communist state.

That state doesn't value each Private's life *per se,* only his wartime potential. That's why its soldiers are more thoroughly trained than U.S. counterparts. Some of the most helpful instruction for USMC personnel during WWII came from their Staff Non-Commissioned Officers (SNCOs) before the officers got to work.

Previous Hints from the East

In the daytime, the WWII German squad would go after an Allied bunker through a double envelopment with a central MG as their base of fire. The flanking elements used smoke to cross open ground and quickly advanced after shooting. That way, any return fire was directed at non-occupied areas. The element most closely approaching the target would conduct the final assault with smoke and concussion grenades only.[6]

But, that smoke and movement also alerted the Allied gun crew to impending trouble. Its total surprise would not come until the enemy night attacks of the Vietnam era. Then, the only noise was what sounded like mortar impacts. If not killed outright by an RPG round, U.S. bunker occupants were too shaken up to properly function. Their emplacement wasn't initially captured, but so thoroughly surrounded as to offer little resistance (just as the American Indians liked to do [7]). The NVA made no attempt to segment the objective and annihilate every enclave. Every frontline assaulter simply looked for ways to help while making his way to the rear of the objective. In the movie, "We Were Soldiers Once and Young," a lone NVA Private tries to kill the U.S. commander single-handedly.[8] By continuing on through the enemy camp, he and his buddies (like the Chinese before them) had kept defenders on the other side of the perimeter from turning their weapons inward. In a hilltop setting, such a move works quite well against too thorough a sweep.

What This Means to Ideal MW Technique

Minimum range for an RPG-7 or M153 SMAW (Shoulder-Launched Multipurpose Assault Weapon) is about 80 feet.[9] Any bunker-busting munition that arms at a greater distance would have limited utility in the fight for a compact bunker complex. With an arming limit of 40 feet, mechanical grenade throwers would have more luck penetrating an aperture. But, most promising would be the baseball players who had regularly tossed hand grenades for distance and accuracy. That, after first crawling through the micro-terrain, will remain the best way to surprise strongpoint occupants.

Artillery shells or aerial missiles can't neutralize such a target because its exact location can't be determined without getting to within a few feet of its gun port. On Okinawa, the Japanese placed their reverse slope defenses just below the top of each hill so that the hill's natural curvature prevented crest-located bazooka men from easily acquiring a target.[10]

9 With Regard to Gaps and Firesacks

- Does attacking through an apparent gap always work?
- What can the space between foe strongpoints turn into?

Unobstructed passageway can lead to a deadly trap.

(Source: FM 7-11B1/2 (1978), p. 2-II-B-3.3.)

Maneuver Warriors Like to Attack Where the Enemy Isn't

Any apparent weak spot in the opponent's protective array is called a "gap." But, U.S. forces more often exploit such things on a grandiose scale than through the microterrain of an enemy perimeter. While more mechanized than the Marines on Okinawa during WWII, the U.S. Army continually looked for openings in the island's first line of defense. The 27th Division bypassed Item Pocket and Kakazu Ridge to more easily capture Machinato Airfield. After enduring a 320mm mortar barrage at the Buzz Bomb Bowl near Machinato Inlet, the 27th came back to engage Item Pocket and

Kakazu Ridge. A mechanized American force had already dared to pass through the small valley between Kakazu and Nishibaru Ridges.

> In the [subsequent] fighting around Kakazu Ridge, the Japanese had separated the [U.S. Army] tanks from their supporting infantry by fire, then knocked off 22 of 30 Shermans with everything from 47mm guns to hand-delivered satchel charges.[1]
>
> — *The Final Campaign,* by Alexander
> History and Museums Division, HQMC, 1996

Through the MW concepts of "reverse-slope defense" and "firesack," the defenders of Okinawa had made ample provision for any U.S. penetration of their lines. Their soft, gradually receding strongpoint matrix made any deep interdiction easier to cut off and annihilate.

Near the top of almost every Okinawan hill's reverse face had been a fully bunkered perimeter. It could accomplish three things: (1) plaster the hill front with preregistered mortar rounds; (2) mow down attackers as they crossed the crest; and (3) take all areas to the side and rear of the hill under standoff fire. The latter capability was the most interesting from a retreat combat standpoint. Not only did it provide mutual support to other hills, but turned the lowland between them into firesacks. These multi-sided ambushes made any attempt at an armored envelopment very risky.

At the main Shuri defense line, "tanks were unable to creep around the west slope of Sugar Loaf because of antitank fire from several directions." In essence, the Japanese had created kill zones so lethal that no intruder could occupy them without unacceptable losses.[2]

What Constitutes a Firesack in Supposed Front Lines

Since 1917, most Eastern defenders have replaced the traditional string of two-man holes with squad strongpoint belts. To not remind Western attackers of the change, they have retained a continuous barbed wire fence. (See Figure 9.1.)

After breaching that protective wire, Western attackers fre-

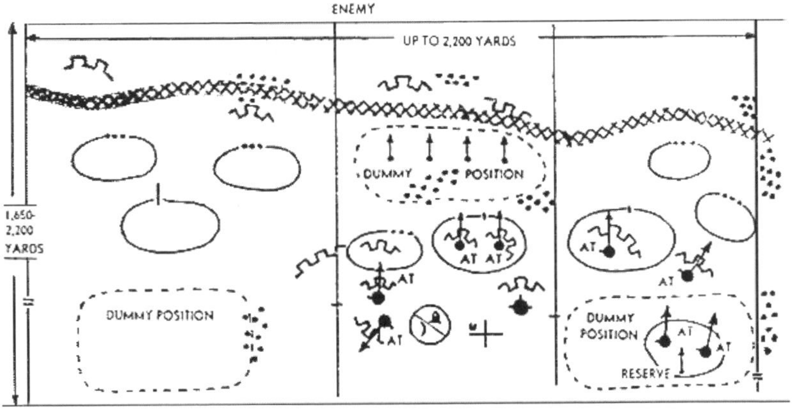

Figure 9.1: WWII Soviets Add Fake Positions to Strongpoint Belt
(Source: *Handbook for U.S.S.R. Military Forces,* TM 30-340 (1945), p. V-47)

quently choose the path of least resistance between those squad strongpoints. (See Figure 9.2.) But, all Eastern squads have their own organic MG. So, GIs frequently find themselves in a multi-sided

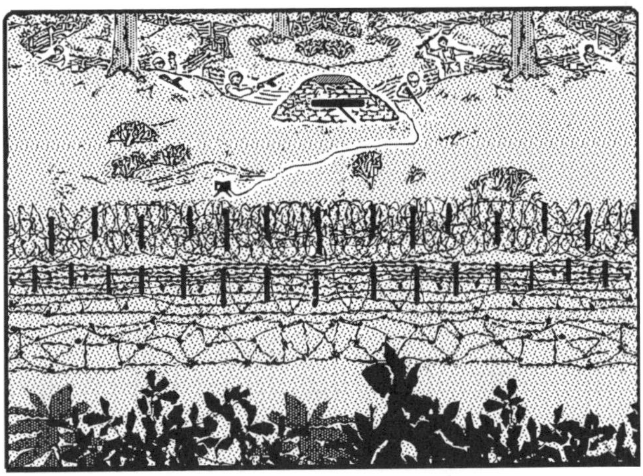

Figure 9.2: GIs Don't Expect a Matrix of Squad Strongpoints
(Sources: FM 7-8 [1984], p. 3-28; MCO P1500.44B, p. 14-18.)

ambush (firesack) along with plenty of land mines and mortar registrations. As on Iwo Jima, they then take fire from ground already seized.

10 Objective Consolidation

- When did America successfully conclude it last conflict?
- What does it take to pacify a huge enemy region?

Overall control only possible through a series of squad outposts.

(Sources: FM 100-5 [1994], p. 39; FM 5-103 [1985], p. 4-5.)

Attack Objective Consolidation Not as Easy as It Sounds

The U.S. military hasn't decisively won a big ground war since WWII, because of problems pacifying a huge area. American forces may have controlled its countryside by the day, but not during the night. Overall control takes more than aerial surveillance. The entire place has to be physically outposted, and all village leaders brought into political consensus.

Unlike the NVA, NKPA, and Iranian Revolutionary Guard, the Pentagon still lacks rifle squads with enough light infantry skill to single-handedly outpost every liberated town and neighborhood.

Only during the first part of the Vietnam War were the squads of the USMC's Combined Action Platoon (CAP) successful at reinforcing village cadre. Then, they defended that village through roving fireteam ambushes along its periphery.

Just securing the interior of a small attack objective may take some squads firing and moving on line while others advance in column. It took British assaults troop many hours to fully capture a tiny hill in the Falklands in 1982.[1]

The most severe combat always has been conducted at the squad level. So, U.S. squads rehearsing surprise-oriented and terrain adapted techniques during daily PT is undeniably the best way to prepare for WWIII. Unfortunately, most Fleet USMC units practiced the "no-thinking" immediate-action drills of Attrition Warfare until about 1986, and—with the onset of MW—none at all. As a result, such units are not as combat ready as they like to profess. That particular oversight can be most quickly remedied through the SNCO academies.

What a Semi-Autonomous Squad Can Accomplish

A self-sufficient squad can safely operate on an active battlefield without any supporting-arms umbrella. So doing may involve

— FIND/DISRUPT A LARGER ENEMY FORCE BEFORE IT CAN ATTACK.
— HANDLE A MUCH BIGGER OPPONENT IN A CHANCE ENCOUNTER.
— EXTRICATE ITSELF INTACT FROM AN ENEMY AMBUSH.
— AMBUSH, WITH LITTLE RISK, AN ENEMY FORCE OF ANY SIZE.
— PERFORM SHORT-RANGE INFILTRATION OF ENEMY LINES.
— SINGLE-HANDEDLY FORCE A GAP IN ENEMY LINES AT NIGHT.
— SINGLE-HANDEDLY FORCE A GAP IN THOSE LINES DURING DAY.
— DEFEND AGAINST INFILTRATOR, TANK, AND NBC ATTACKS.
— FROM SUCCESSIVE FALL-BACK POSITIONS, DEFEND A CITY BLOCK.
— CLEAR ONE SIDE OF A STREET FASTER THAN FOE THINKS POSSIBLE.

Table 10.1: What a Single Common-Sense MW Squad Can Do

Figure 10.1: There's No Talking on a "World-Class" Patrol
(Source: *FM 7-8* (1984), p. 5-22)

hastily attacking a much larger unit, and then breaking contact. It may require enough counter-tracking ability to elude a determined pursuer. (See Figure 10.1.) Or, it may take busting out of an encirclement. (See Figure 10.2.) Such skills are widely applicable to many MW missions.

What a lone squad trained in common-sense MW can regularly accomplish in close proximity to a large enemy force is truly amazing.[2] (See Table 10.1.) The foe not realizing that an opposition squad is present is what protects that squad.

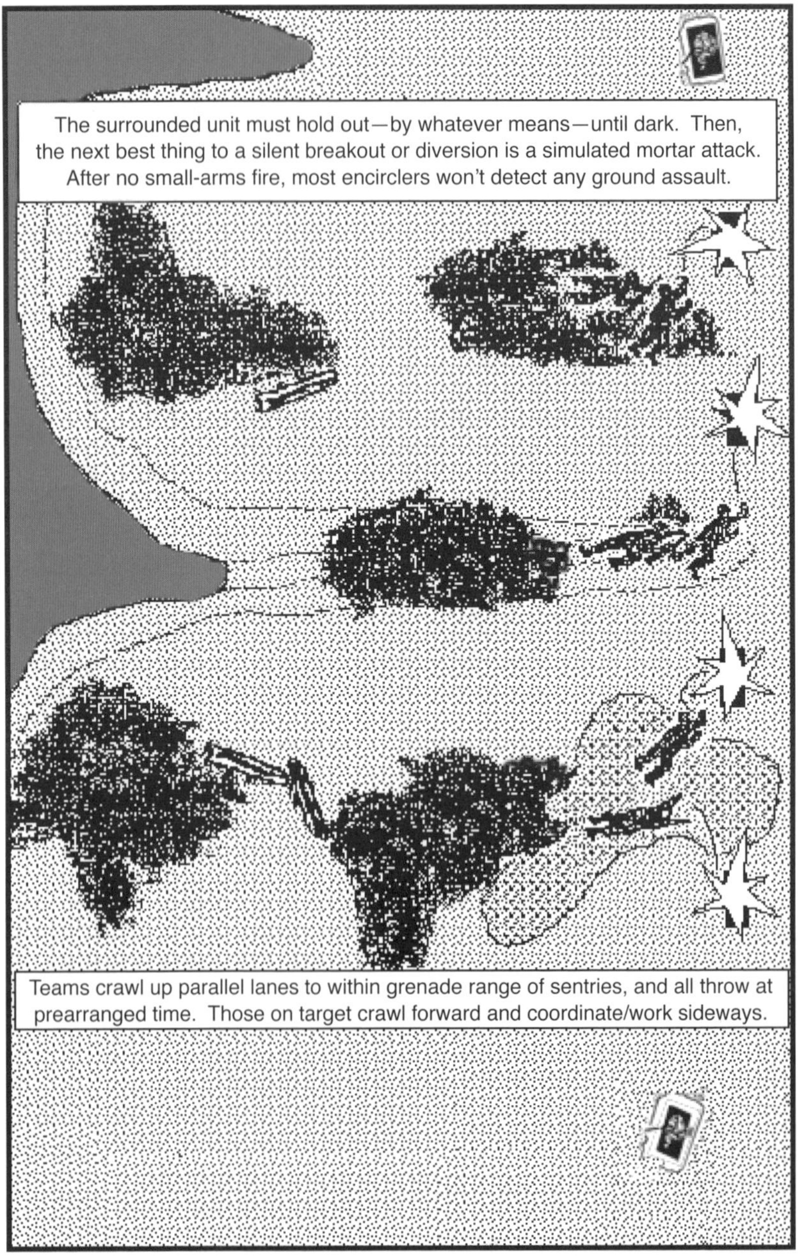

Figure 10.2: Next Best Night Breakout Is a Fake Mortar Attack

(Source: FM 7-70 [1986], p. 420; FM 5-103 [1985], p. 4-6; MCI 03.66a [1986], p. 2-8; FM 23-30 [1988], p. 2-8; FM 7-1183 [1976], p. 2-VII-C-4.4.)

Part Three

School House Procedures

"Reading books is learning, but application is . . . the more important form of learning." — Mao Tse-tung

(Source: "Mao Tse-tung: An Anthology of His Writings," ed. Anne Fremantle (New York: Mentor, 1962), p. 82.)

11 WITH COMMON SENSE AS ONE'S COMPASS

- Which USMC component has the most common sense?
- What is its heritage with regard to combat training?

NCOs have more organizational awareness than other ranks.

(Source: FM 90-5 (1982), p. 2-12.)

Responsibility for the Most Valuable Infantry Training

Prior to 1973, the Gunnery Sergeant conducted most of the individual and small-unit training within each USMC infantry company. FMFM 6-4 still says he is "the principal enlisted assistant to the company commander in supervising the training of the company."[1]

Because Marines of every military occupational specialty (MOS) are obligated to be riflemen, the senior SNCO of every support unit must make sure the infantry skills of all specialists are maintained. But, the skills necessary to fully employ the USMC's new MW doctrine are considerably more complicated than those of Attrition Warfare. Besides knowing how to shoot and follow orders, the modern-day Marine has to be at least aware of all the micro-basics associated with "shoot, move, and communicate." At a minimum, he must be a good enough light infantryman to accomplish the cornerstones of UW—short-range infiltration and secretive ambushing. That means all (Non-Commissioned Officers) NCOs will require refresher infantry training every time they return to the regional SNCO academy. And, almost all such training must be of a practical-application nature. Their higher degree of organizational awareness (common sense) has already created the proper MW mindset. They are, and always have been, the technicians of actual combat.

All Tactical Instruction Must Be Experimental

Tactics are the maneuvers that mostly occur at the small-unit level. On Peleliu, Iwo Jima, and Okinawa, most enemy headquarters were in terrain so heavily compartmentalized that only semi-autonomous fireteams could reduce it through mutual cooperation (*gung ho* infantry action). Tanks and heavy artillery were of no help. But, this simple truth—that the infantry Private is the most important person in battle—hasn't survived any U.S. war. That's because standardized procedure and following orders seemed more conducive to large-unit maneuvers. And, broad brush strokes are what top-down Western militaries prefer. The Marine NCO's amazing degree of insight into such matters originated from having to occupy a tiny cubicle at one corner of a large squad bay filled with heart breakers and life takers. He couldn't help but learn how they collectively felt about things.

As soon as any foe realizes which U.S. tactical procedures are being used, they become far too predictable. So, every modern-day American rifle company needs its own unique set of tactical techniques (squad-sized and below moves). The days of division formations across a wide frontage are over. Nothing larger than a well-dispersed company will be safe from aerial destruction. And, each set of small-unit methods must be continually improved and adapted to local conditions. This is most easily accomplished through trial and error.

Fewer casualties are the best measure of tactical progress.

So, all training should involve competition with regard to speed, stealth, or deception. That competition can be between separate individuals or one against his previous attempt. The whole idea is to better surprise the foe or bypass him altogether. Closing with and destroying him is no longer the primary goal. Where MILES gear is not available, the three-second sight pictures of a rubber rifle will suffice, as can traces of flour grenade impact. Then, by practicing his parent unit's movement techniques during daily physical training (PT), the member of every Marine company can maintain enough combat proficiency to perform the offensive or defensive missions of MW.

The Last MW-Oriented School for Enlisted Personnel

Whereas boot camp had taught every Marine unit discipline, Lt. Col. Carlson tried to enhance the self-discipline, individual skill, and personal initiative of his Maoist Raiders. So doing took a slightly different approach to training. No longer was it strictly group oriented. Yet, small group dynamics were still stressed. The student body was divided up into squad-sized contingents that did almost everything together.

Carlson's initial emphasis was on how to win every one-on-one encounter. He wanted only those Marines who were prepared to kill or be killed at close range with a knife. Then, he gave them advanced training in close combat and endurance. His troops learned to watch for an attack against their person from any direction and then—through some unique ruse—instantly gain the upper hand against an onrushing assailant. Their guidance was "always to expect the unexpected."[2]

At some point, "Red Mike" Edson also realized that his junior enlisted personnel needed more self-confidence to do well at short-range combat. As had Carlson, he found free-play to be the most realistic way of training both them and their tiny groups.

> If I had to train my regiment over again, I would stress small group training and the training of the individual....
>
> Our basic training is all right....In your training put your time and emphasis on the squad and platoon rather than on the company, battalion and regiment.

> In your scouting and patrolling,...have the men work against each other. Same thing for squads and platoons in their problems....
>
> ...With proper training, our Americans are better [than the Japanese], as our people can think better as individuals. Encourage your individuals and bring them out.[3]
>
> — Col. Merritt A. Edson USMC

While the first Raider preparation had been at Jacques Farm, a satellite facility of Camp Elliott, the "Raider Training Center" was eventually at Camp Pendleton. There occurred eight hard weeks of instruction. Its subjects ranged from Guerrilla Warfare to individual cookery. The culmination was a week-long field problem in which students alternately made up the main body of friendlies and two opposing groups of guerrillas. There was also rubber boat training. The rest of the curriculum dealt mostly with traditional methods: (1) marksmanship; (2) scouting; (3) patrolling; (4) personal conditioning; and (5) individual combat. Among the group skills were vaulting barbed wire, crossing rivers on two-rope bridges, crawling through low obstacles, and parallel-column road hiking. Among the individual skills were knife fighting,[4] wall scaling, barbed-wire crushing, and invisible-obstacle vaulting (like one would do over a stream of MG bullets).[5]

Particularly productive of initiative was each Raider's chance to trick an onrushing assailant during self-defense training. This had come through Lt.Col. Roosevelt at Raider School from Carlson.

12 Grenade Throwing for Distance & Accuracy

- Why does bunker destruction still require a hand grenade?
- How many little pineapples were used on Okinawa?

Most enemy gun ports are only recognizable up close.

(Source: "Closing In: Marines in the Seizure of Iwo Jima," by J.H. Alexander, Marines in WWII Commemorative Series, HQMC, 1994, drawing by Charles H. Waterhouse, p. 31)

A Modern War Still Requires Grenade Work

Nowhere in the Pacific late in WWII did the heavy American barrage make much difference.[1] The exact location of the enemy's automatic-weapons was nearly impossible to ascertain from a distance. At Peleliu, two Yanks needed white-phosphorus grenades to crawl close enough to fire a rifle grenade through the bunker aperture of the Point's infamous 47mm cannon.[2] On Iwo Jima, one or two low-ranking volunteers had to squirm around each enemy strongpoint to drop a "frag" grenade into every crevice on a more direct route home.[3]

By the time Okinawa rolled around, not much "blind 'em, burn 'em, and blast 'em" technique from the Tarawa era was still being practiced in its entirety.[4] The most dangerous parts of that bunker-busting routine had been replaced by a bazooka blast or smoke screen (to facilitate the approach), and hand grenade through the gun port (instead of a satchel charge).

Then, something happened on Okinawa to which U.S. historians should have attached more importance. Almost every confrontation between sides quickly deteriorated into a hand grenade duel. Because the Japanese had also preregistered the same piece of ground for indirect fire, they had little trouble winning the initial engagement.

Normally, such a flurry of tiny pineapples would take place as the Americans crossed the very top of an assault objective, and their opponent counterattacked from its rearward face. But, this identical sequence of events occurred so often during the 82-day battle that the Japanese may have taken their cue from the Finns (through the Germans). By doctrine, Finnish troops will only assault from within grenade range of their opponent (20-30 meters away), and after every friendly has tossed a grenade.[5] The counterattacking Japanese had emerged from their reverse-slope positions at about this same distance—right after their pineapple shrapnel had completely engulfed any crest-crossing GIs.

At that point, distance and accuracy of hand grenade became more important than any caliber or degree of precision for a new cannon or missile launcher. Yet, modern-day U.S. infantrymen will be lucky to have thrown a single grenade (live or otherwise) over the course of a career.

A Feasible Drill for Accuracy

The circular net used by baseball pitchers would best help one's grenade throwing accuracy. But, only in the woods could a stand-up toss be accomplished in relative safety from behind a tree. (See Figure 12.1.) Elsewhere, the prone position provides the most cover. Then, the cross-body toss launch from a reclining position can be successful against an enemy gun pit with no trees in the way. The grenade will bounce a little and then roll into the hole.

All training should be fast-moving and fun. The sound made by

Figure 12.1: Upright Toss Has Best Result in the Woods
(Source: FM 7-8 [Aug. 1984], p. B-3)

a practice grenade hitting a 55-gallon drum allows for an instant and accurate count of actual hits. Valuable time will also be saved retrieving the little missiles. (See Figures 12.2 and 12.3.)

Figure 12.2: Only Normally Safe Throw Is from the Prone Position
(Source: FM 23-30 [Dec. 1988], p. 2-8)

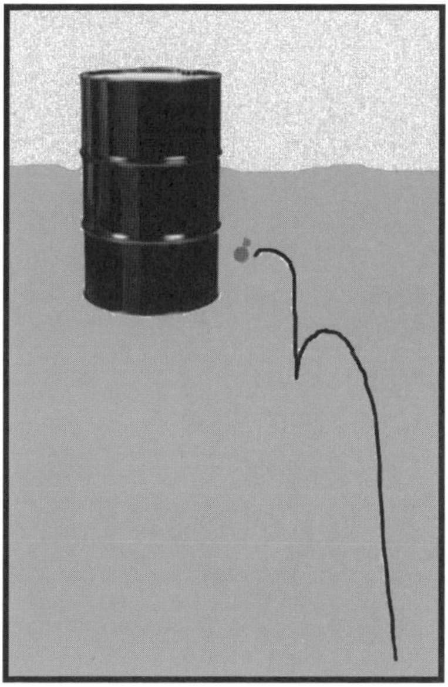

Figure 12.3: Easiest Way to Count Hits in Training
(Source: FM 23-30 [Dec. 1988], p. I-30)

13 Crawling for Distance and Secrecy

- Can prone movement escape modern surveillance devices?
- How far and fast should an infantryman be able to crawl?

Safely attacking a foe MG almost always involves crawling.

(Source: MCRP 3-02H [1999], fig. 1-2)

How Crawling Became the Key to Post-MG Tactics

Entire generations of European infantrymen had already succumbed to the recently invented MG in WWI, when 30,000 U.S. Doughboys met the same fate during its last two months. Their mistake was trying to move upright through the fields of fire of hidden automatic weapons. Such a killing machine must be approached more carefully than that. It's most safely done by crawling to outflank it through the microterrain. But, the participants of a large on-line infantry formation don't have that luxury. Their primary responsibility is keeping up with the people to their left and right.

Since 1917, a German-Stormtrooper-like squad has been able to secretly approach any Allied defense line through old communication trenches, bangalore its barbed wire during a precision bombardment, and then crawl in throwing concussion grenades as the big shells shifted rearward. They did so where no Allied MG was located, so that such a gun could be assailed from a flank. At that point in history, losing many U.S. troops to MG fire became unnecessary. Just by temporarily dispersing, any maneuver element could have escaped long-range destruction. But, the Western armies could not—or would not—create skilled enough squads to operate alone. The rest is history, and not a very happy one at that.

Long Distance Crawling Takes Extra Muscles and Stamina

On Guadalcanal during WWII, the Japanese were to conduct entire patrols near U.S. lines in the prone position.[1] A Western soldier could develop the same capability through enough practice. In the modern-day U.S. military, such training is not considered harassment if an instructor participates. Normally, it takes the form of back-and-forth crawl racing between three separate units. (See Figure 13.1.)

Microterrain Often Provides the Only Safe Route

Cross-country crawl racing for time can usefully ensue. Such a route would include watershed ditches, depression between mounds, swamps, tree-falls, and the like. (See Figures 13.2 through 13.7.) Across any open space at night, movement slow enough to be imperceptible in low light will also work. (See Figure 13.8.) And last, but not least, on-line assaults should be be practiced while crawling. (See Figure 13.9.)

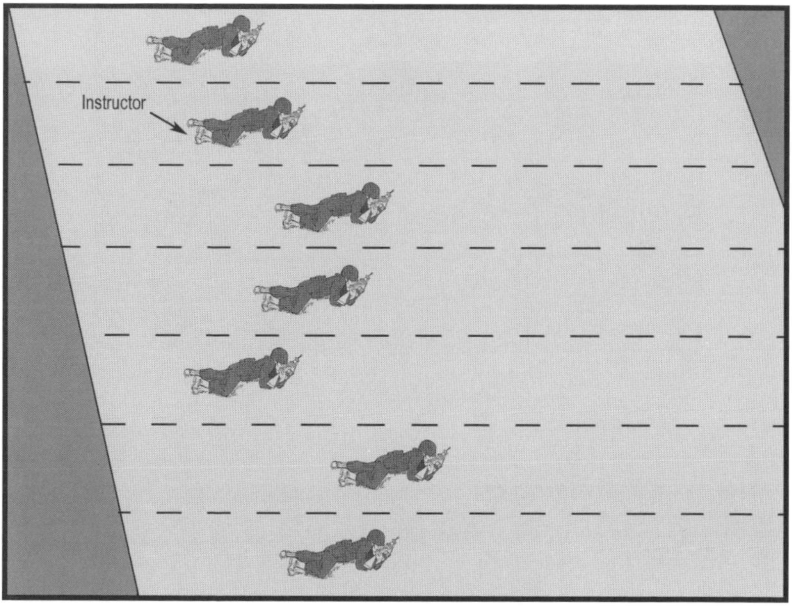

Figure 13.1: Crawl Racing
(Source: FM 100-5 [1994], p. 56)

THE SURFACE OF THE EARTH IS COVERED WITH TINY WATERSHED DITCHES. BECAUSE MOST DEFENSES ARE ON ELEVATED GROUND, THEY HAVE DITCHES LEADING INTO THEIR INTERIOR.

Figure 13.2: Rainwater Runoff Indentations Provide Cover
(Source: FM 5-103 [1985], p. 4-4)

Figure 13.3: Fallen Trees Offer Lengthwise Cover
(Sources: FM 7-11B1/2 [1978], p. 2-II-A-8.2; FM 7-11B3 [1976], p. 2-VII-C-4.4)

Figure 13.4: Spaces Between Bushes Give Only Concealment
(Source: FM 5-103 [1985], pp. 5-2, D-9)

Figure 13.5: Low-Ground Has Its Benefits
(Source: MCO P1500.44B, p. 12-63)

Figure 13.6: The Foe Is Easier to Spot from Below
(Sources: FM 7-11B1/2 (1978), p. 2-III-E-8.2; FM 5-103 (1985), p. 4-4; FM 22-100 (1983), p. 22)

Figure 13.7: Damp Areas Are Seldom Occupied
(Source: FM 21-76 [1970], p. 43)

Figure 13.8: Very Slow Movement Is Hard to Notice in a Clearing
(Sources: FMFM 0-1 [1979], p. 7-7; FM 21-75 [1967] p. 2-II-A-4.2)

Figure 13.9: On-Line Assaults Need Not Be Upright
(Source: "Army/Marine Clipart," U.S. Air University, from this url: www.au.af.mil/au/awc/awcgate/cliparmy.htm)

14 Daily Boots and Rifles Fun Run

- What can combat physical fitness training (PT) accomplish?
- Is there an appropriate training aid?

All stamina building should be battlefield oriented.

(Source: Air Univ., Army/Marine Clipart, image designator "1-4b.gif," retrieved from www.au.af.mil/au/awc/awcgate/cliparmy.htm)

"Indian File" Squad Runs Will Simulate MW

U.S. travel speed greater than an adversary expects will disorient him. It makes him less able to adapt to changes of circumstance.

> Physical speed, moving more miles per hour, is a powerful weapon in itself. On our approach to the enemy, speed in movement reduces his reaction time. When we are going through him or around him, it changes the situation faster than he can react. Once we are past him, it makes his re-

> action irrelevant. In all three cases, speed impacts on the enemy, especially his mind, causing fear, indecision, and helplessness. Remember, attacking the enemy's mind is an central tenet of maneuver warfare.[1]
>
> — "Tactics," *MCDP 1-3*

The quickest moving infantry element capable of surviving on its own is the squad-sized column or file. The British formed somewhat larger "flying columns" toward the end of WWI to move deeper into enemy territory after capturing a fortified trench by on-line assault.[2] But, only lone German Stormtrooper squads could penetrate linear Allied defenses in column formation. They did so through a simulated artillery attack. Then, Japanese squads were to separately attack U.S. lines single file during WWII.[3] So would Chinese squads in Korea,[4] and NVA squads in Vietnam. In the NVA version, an RPG man led the way.[5] (Look back at this book's cover.)

The Subic Bay Precedent

Around 1989, a student at the USMC's East Coast Platoon Sergeants School told an instructor of a technique he had used to patrol the Subic Bay Naval Base boundary in the Philippines. While dark-clothed sentries sat in occasional trees, U.S. patrols jogged single file along the fenceline trail. There were guerrilla factions in the region, but none had been very active.

For this Marine unit, suddenly appearing had been thought to better surprise an adversary than a stealthy approach. Its methods were a perfect example of MW, before that alternative way of fighting had become USMC doctrine. "Elevated" or quickly moving sentries could monitor a broader defense zone.

PT Is the Perfect Time to Practice/Refine Squad Maneuvers

As discussed in Chapter 4, single-file formations are also quite valuable to approach marches and patrolling. While crawling along the various depressions in the ground, those movements become difficult to spot—even in a "high-tech" environment.

There are no rear areas in modern war. Every tiny enclave must be properly protected. This takes foolproof perimeter security and expert contact patrolling. Defensive patrols must be able to engage an opposition force of any size without taking any casualties. So doing disrupts its momentum long enough for the parent unit to mount a successful defense. Such a contact can only be accomplished through a well-rehearsed "fire and movement" drill. One of the safest is where an on-line squad closes with the foe through random and well-protected individual rushes. Each rusher enjoys the covering fire of his buddy team partner. Then, the squad either completes its assault through a forward-element grenade toss or a "creeping" mortar barrage. Any retrograde operation is largely the same events in reverse order. Though almost never practiced by the U.S. military, such a move is perfectly legitimate in MW.

Things for the Movement Column to Rehearse

Each squad can practice the following as it makes its way around a clearly marked circuit: (1) reconnaissance pull (sending out scouts to reconnoiter the best route); (2) immediate ambush (moving sideways into good cover); (3) on-line investigation (a staggered advance without any shooting; (4) hasty on-line assault ("fire and movement" followed by a grenade toss); and (5) retrograde operation (leapfrogging backwards).

All "fire and movement" will be accomplished through upright individual rushes that never last more than three seconds and always end up with a short crawl to some defiladed position. Such a position might be a tree, stump, rock, mound, shell hole, or other low spot. No ordinary bush will qualify. Where not enough pieces of cover exist, the advancing individual can crawl the whole way. All teams must remain fairly well aligned.

Side Benefits of an MW Physical Fitness Program

Every non-infantry Marine must still maintain rifleman proficiency. In a training environment now awash with virtual reality, so doing takes additional "woods" skills. Carrying a rubble rifle during PT helps to establish them.

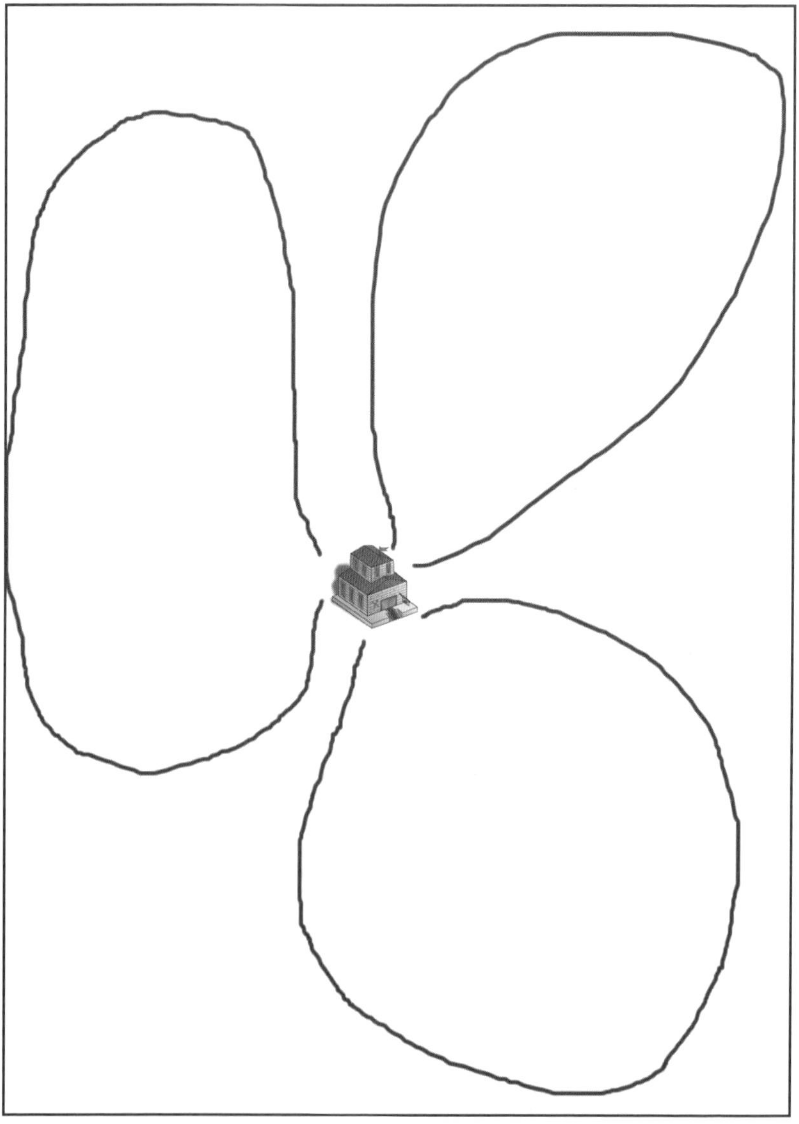

Figure 14.1: Clover Leaf Circuits from the School House
(Source: PublicDomainVectors.org, with image designator "nicubunu_RPG_map_symbols_Townhall.jpg.")

The infantry MOS is one of the most complicated career fields. While the Corps has mostly counted on heavy infantrymen in the past (those who know their weapon and how to follow orders), there are many light-infantry skills. In an under-equipped rear-area unit, the latter prove most useful. They will also permit frontline GIs to befuddle the latest target acquisition devices.

There two types of combat PT—that for Attrition Warfare and that for MW. The former often entails carrying a certain load over a designated distance. In the old days it was a wounded Marine. Now, it's a simulated load of extra ammunition. But, most MW missions are accomplished through more maneuver than firepower, and don't require many bullets.

MW physical fitness will focus on personal initiative, agility, and speed. Because enemy fire is always possible in a war zone, not many upright skills should be practiced in a peacetime open field. More appropriate would be crawling or grenade-throwing from the prone position. Moving forward for less than three seconds, dropping to the ground, and wriggling forward a short distance might also be appropriate. The MW goal is to out-think the foe.

Initial Conduct of the Boots and Rifles Fun Run

Now that dismounted U.S. grunts face drone surveillance, the similarity between squad column run and foot patrol/approach march becomes more compelling. The 15 or so members of each student file are told to imagine sniper fire while moving around a one-mile loop near their school house. To cross an open area, they must dash three at a time or crawl as a group. Elsewhere, they shall mostly move in a zigzag pattern along linear "cover" (like tree lines, building fronts, fence lines, or ditches. By continuing to perform this exercise daily, all school graduates (including the Fleet infantrymen) will always be ready for active combat.

To make such a journey fun, it must be completed as quickly as possible, with an extra minute added for each execution error. Every same-circuit squad will be in competition with the others, and the fastest time be best. Daily winners should receive some compensation—like no cleanup duty.

With three separate trails marked by different colored engineer tape, any number of student files could deploy at prescribed intervals. (See Figure 14.1.) An instructor would be stationed along each

Figure 14.2: Point Scout Chooses the Main Body's Exact Route
(Source: FM 21-75 [1967], p. 8)

route to monitor required events and periodically assign a new one. Once the basic passage is mastered, various enemy contacts can be added (as listed below). At first, the same encounter is practiced in several settings, then other contacts. Every student should serve as navigational scout/road guard and squad leader.

Figure 14.3: Main Body Can Run Between Stops
(Source: Air Univ., Army/Marine Clipart, retrieved from www.au.af.mil/au/awc/awcgate/cliparmy.htm)

Figure 14.4: How Foe Patrol Looks to a Hasty Ambush

The "Recon Pull"

A pair of scouts precedes every squad column. The main body must periodically stop for proper reconnoitering, but can move as quickly as it wants at other times. It should still follow the linear obstacles and vegetative screening as much as possible. (See Figures 14.2 and 14.3.) There should be no talking all during every patrol.

The Hasty Ambush

Normally, for an impromptu ambush, the point man signals for the main-body column to face left or right and move farther off the trail they've been following. (See Figure 14.4.) By installing a few claymores and then quickly reeling in the wires, it can hurt an intruder without him ever suspecting an ambush. He instead suspects a minefield or mortar barrage.

The Counter-Ambush

After being ambushed, that part of the main body not in the kill zone crawls back along the trail and then comes on line perpendicular to the ambushers' linear array. Those few friendlies then attack that array from a flank. (See Figure 14.5.)

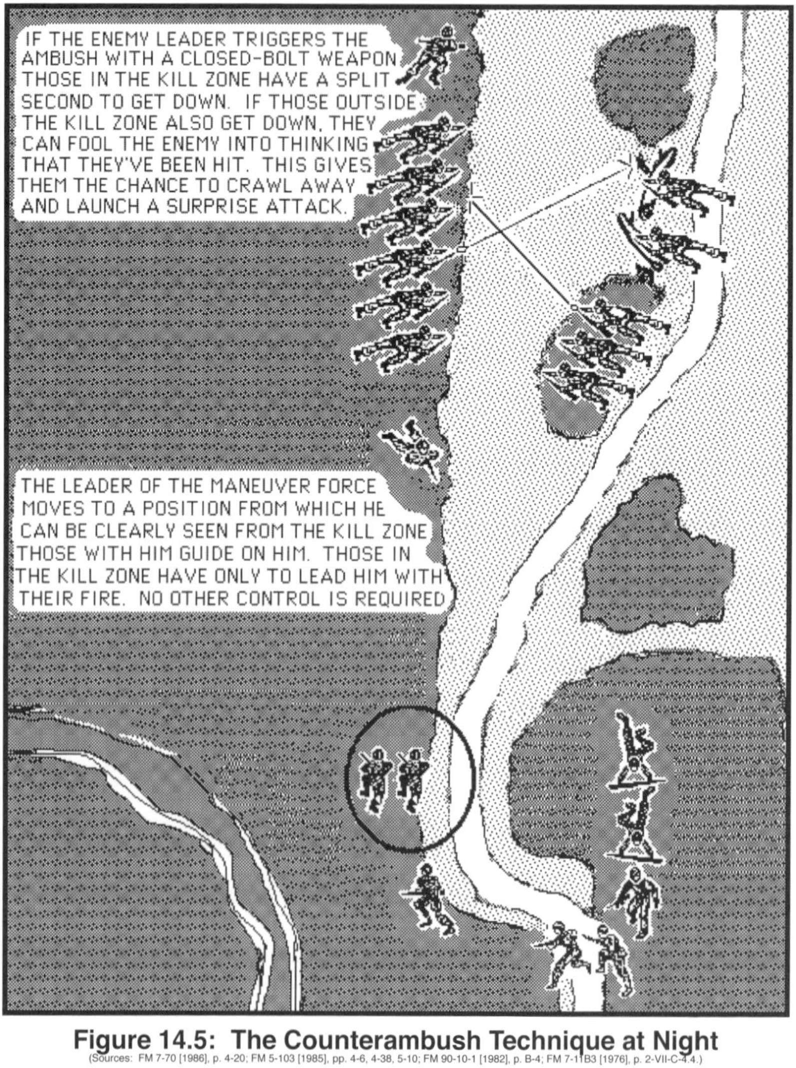

Figure 14.5: The Counterambush Technique at Night
(Sources: FM 7-70 [1986], p. 4-20; FM 5-103 [1985], pp. 4-6, 4-38, 5-10; FM 90-10-1 [1982], p. B-4; FM 7-11B3 [1976], p. 2-VII-C-4.4.)

Figure 14.6: File Alters Direction and Faces to Become a Line
(Sources: FM 22-100 [1983], p. 185; FM 90-10-1 [1982], pp. B-3, B-4)

A Sideways Investigation

Whenever deemed appropriate, the column can face to one side and move laterally on line to look for enemy sign or disguise its own. (Look back at Figures 3.2 and 3.3.)

Chance Contact

Sometimes, everyone in a fairly straight string can stop and then move sideways on line toward an approaching enemy. Other times, they must turn an imaginary corner, move directly forward a short distance, and then all face the foe. (See Figure 14.6.)

The Fire and Movement Drill

This way of closing with a foe can be done upright and without any shooting until spotted by the quarry. Then, every individual rush must be random and carefully conducted. That takes each rusher being up only three seconds and then changing location before reappearing. Crawling forward to some form of cover is better than rolling to one side. (See Figures 14.7 through 14.10.) Nearby instructors should advise the offending squad of any errors before it moves on.

Figure 14.7: Staying Upright Less Than Three Seconds Is Crucial
(Sources: FM 22-100 [1983], p. 84; FM 7-70 [1986], p. D-24)

Figure 14.8: So Is Changing Location After "Hitting the Deck"
(Source: FM 21-75 [1967], pp. 26, 77)

Figure 14.9: Squad Covers & Moves While Leader Calls for Mortars
(Sources: MCI 03.66a [1986], p. 2-8; FM 22-100 [1983], p. 66; FM 7-70 [1986], p. 4-20; FM 7-11 [1978], p. 2-III-E-8.2; FMFM 6-7 [1989], p. 1-13; FM 5-103 [1985], p. 4-6)

Figure 14.10: Detected Squad Fires & Moves Until Grenade Toss

(Sources: MCI 03.66a [86], pp. 2-8, 2-9; FM 22-100 [83], pp. 66, 84; FM 7-70 [86], p. 4-20; FM 7-11 [78], p. 2-III-E-8.2; FMFM 6-7 [89], p. 1-13; FM 5-103 [85], pp. 4-6,4-38; FM 7-11B3 [76], p. 2-VII-C-4.4)

A Retrograde Operation

Tactically withdrawing from a much larger opponent takes "firing and moving" backwards while seldom standing upright. (See Figure 14.11.) It's most easily accomplished by one or two fireteams crawling rearward through the defiladed folds in the ground, while those remaining continue to shoot at the enemy. Poorly covered moves should be screened with grenade smoke. Each burst of suppressive bullets must be quickly followed by a change of location. Then, all return fire has no effect.

Figure 14.11: Most Rearward Movement While Prone
(Source: FM 5-12B3 [1977], p. 2-308.)

Road Crossing

For a Yank rifleman on patrol, continually stopping at every small opening can be demoralizing. Such openings are only "caution areas." If they can't be bypassed, they can be quickly and safely transited through the "bump method." Before crossing, each buddy team waits for the next two men in column to reach the edge of the opening, face outboard, and cover the exposed movement. (See figure 14.12.)

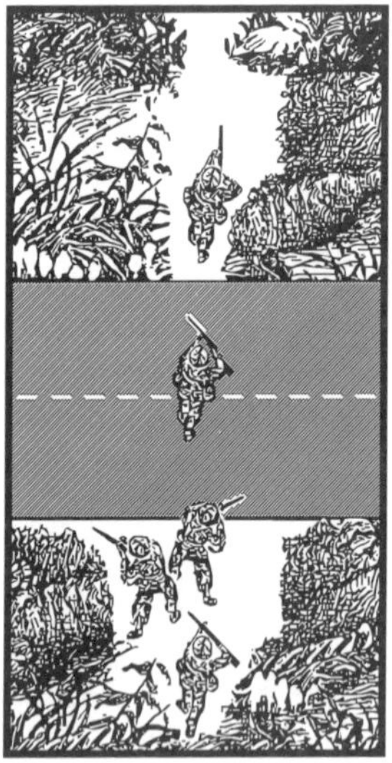

Figure 14.12: The "Bump Method" Will Suffice at Roads
(Source: FM 5-103 (1985), pp. 4-4, 4-6)

15 Stalking & Short-Range Infiltration Course

- What's the value of sneaking up on someone?
- How hard is it to move unnoticed between sentries?

Crawling between watchstanders to enter an enemy camp.

(Sources: FMFM1-3B [1981], p. 4-9; FM 5-103 [1985], p. 4-4)

Stalking a Sentry Takes an Intervening Obstacle

Any piece of screening or solid cover between a stalker and his target will keep that stalker from being noticed during the day. Bushes, rocks, stumps, trees, and mounds all qualify. (See Figures 15.1 and 15.2.) Dressing like a bush and crossing open areas more slowly than perceptible to the naked eye can work at night where no thermal imaging is present. (See Figures 15.3 through 15.5.) In Figures 15.6 and 15.7, a combination of stalking, sniping, and deception allows a friendly flying column to destroy an enemy CP without revealing its own presence on the battlefield.

Figure 15.1: Stalker Keeps Object Between Himself and Quarry
(Source: "Podgotovka Razvegchika: Sistema Spetsnaza GRU," © 1998 by A.E. Taras and F.D. Zanuz, p. 373)

Figure 15.2: Any Mound or Rock Will Obscure the Stalker
(Source: Courtesy of Sorman Information and Media, illustration by Wolfgang Bartsch, from "SoldF: Soldaten i falt," © 2001 Forsvarsmakten, Stockholm, p. 220)

Figure 15.3: Open Areas Less Risky After Dressing Like a Bush
(Source: FM 21-75 [1967], p. 67)

The Stalking Course

Any tree-covered vacant lot can be turned into a stalking course. Three units at either end take turns stalking each other. The acting defender keeps score on how many opponents had can spot beyond grenade range.

The Infiltration Course

After donning MILES gear some night, half of the class tries to slip between the perimeter fighting holes of the other half. Flour grenades can be used instead of MILES to record infiltrator discoveries.

Short-Range Infiltration Is the Hallmark of UW

Because U.S. forces usually attack *en masse* during the day, most GIs consider an American sapper attack too risky. It's not. Where the quarry's fields of fire have been partially cleared or sentries put on reduced alert, U.S. infiltrators could sneak into an enemy camp at noon. If Eastern Hemisphere troops can do something, so can their Western counterparts. The former have no more patience. To

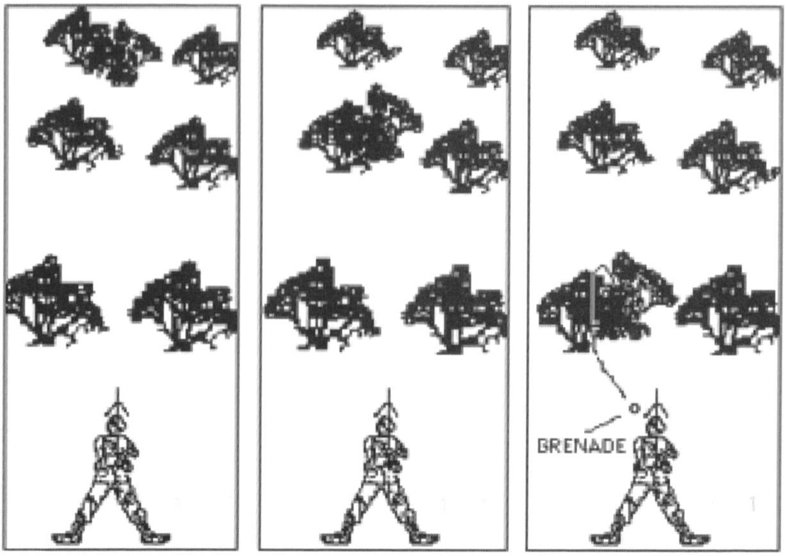

Figure 15.4: Most Sentries Don't Expect a Change in Vegetation
(Source: FM 7-11B1/2 [1978], p. 2-II-E-8.2; FM 7-8 [1984], p. 3-28)

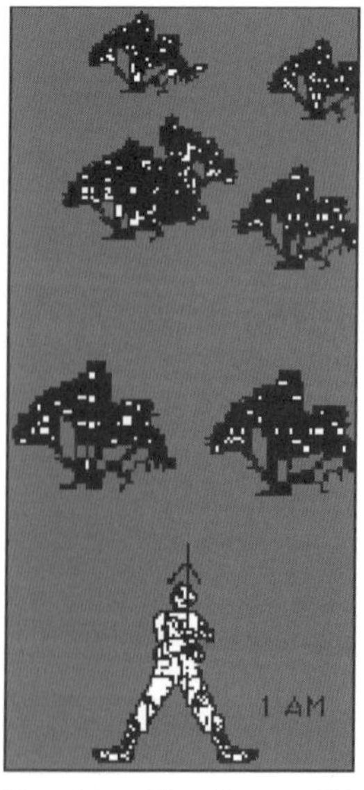

Figure 15.5: Various Bushes Harder to Distinguish After Dark
(Sources: FM 7-11B1/2 [1978], p. 2-II-E-8.2; FM 7-8 [1984], p. 3-28)

develop real patience, one has to spend time at the very bottom of a overly tall Western bureaucracy. Below is a rare account of how "natural" such an attack can feel to an ordinary soldier.

> Once [in the Mekong Delta in 1965] I got into the middle of Cai Be [Army of the Republic of Vietnam (ARVN)] post where the district chief's office was. It was fifteen days prior to the attack and takeover of the post by our battalion. I was accompanied by two comrades armed with submachine guns to protect me in case my presence was discovered while I was nearing the post entrance. I was then wearing pants only and had in my belt a pair of pincers, a knife, and a grenade. At one hundred meters from the post I started crawling and quietly approached the post entrance with the two comrades following me. At twenty meters from the post, my comrades halted while I crawled on. At the post entrance there was a barbed-wire barricade on which hung two grenades. Behind the barricade stood a guard. I made my way between the barricade and the stakes holding up the barbed wire fence. I waited in the dark for the moment when the guard lit his

Figure 15.6: Sentry Can't See Right After Cloud Covers Moon
(Source: FM 7-11B1/2 [1978], p. 2-II-A-4.2)

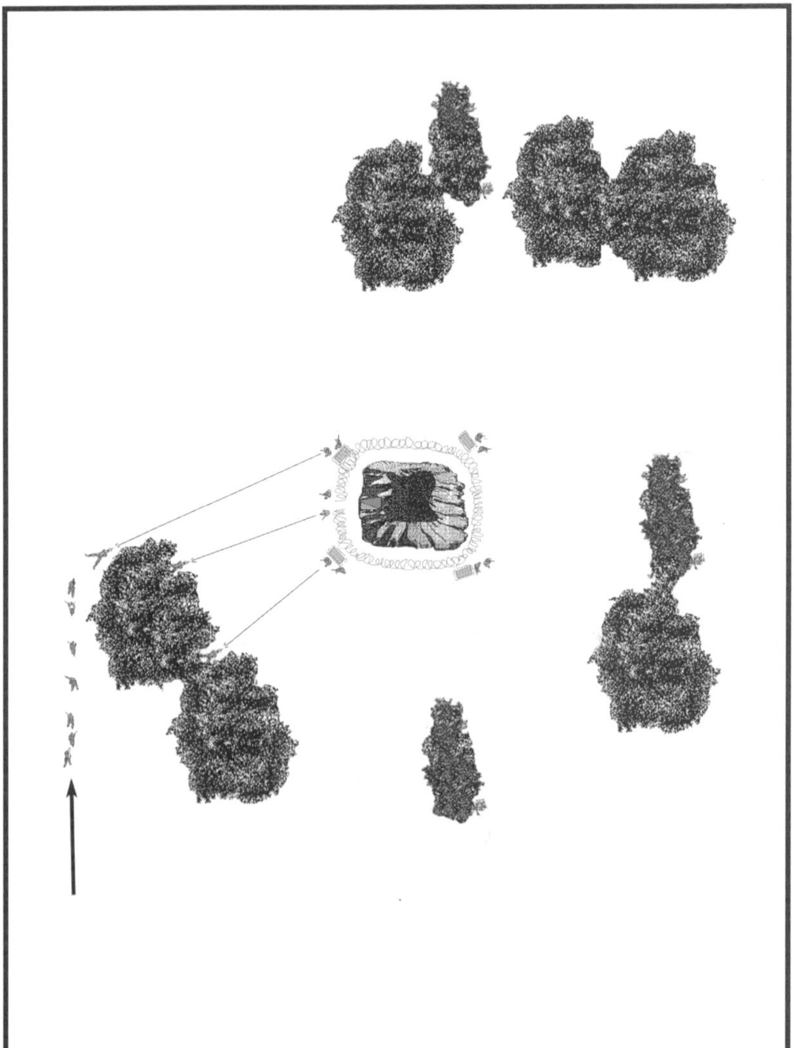

Figure 15.7: How Flying Column Deals with a Command Bunker
(Source: *The Last Hundred Yrds*, Posterity Press, 1997, © 1994, 1995, 1996, 1998 by H.J. Poole, figs. 18.7, 19.3, 20.9)

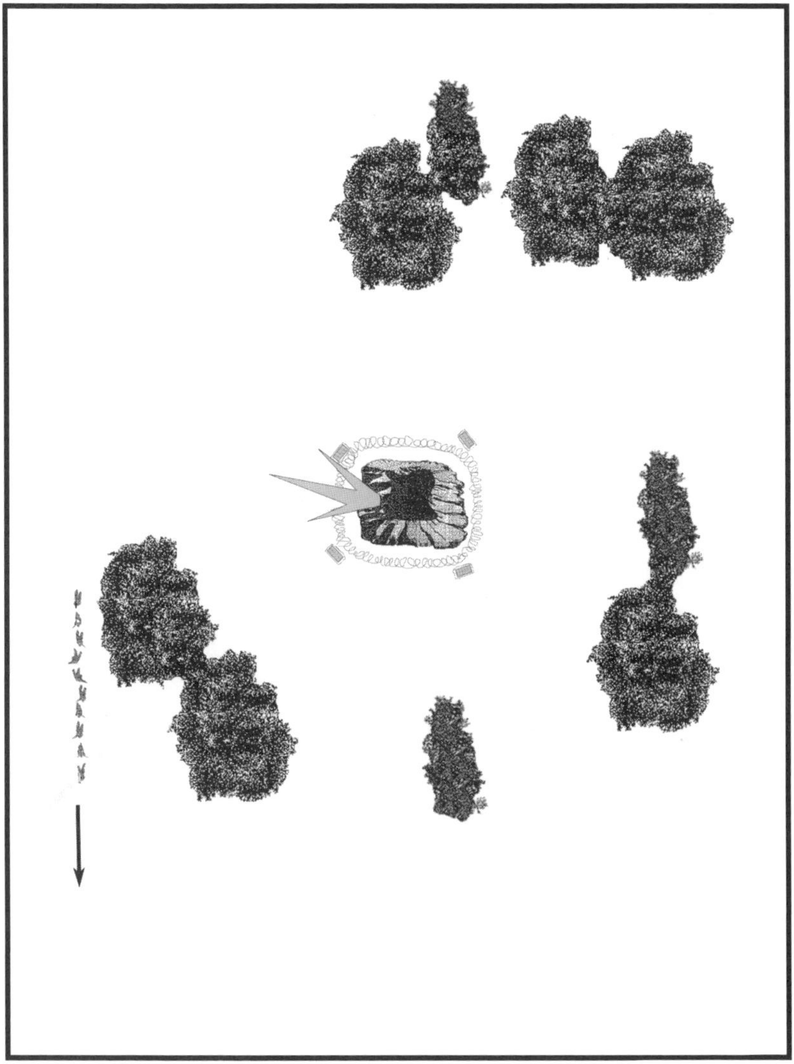

Figure 15.8: Bunker Blast Made to Look Like a Lucky Mortar Hit
(Source: *The Last Hundred Yards*, Posterity Press, 1997, © 1994, 1995, 1996, 1998 by H.J. Poole, figs. 18.7, 19.3, 20.9)

> cigarette. I passed two meters away from him and sneaked through the entrance. On that occasion I was unable to find out where the munitions depot was but I did discover the positions of two machine guns and the radio room. I got out at the back of the post by cutting my way through the barbed wire.[1]
>
> — Reconnaissance sapper
> 261st Viet Cong Battalion

Short-range infiltration is safer than other forms of infantry assault because only one team out of many must enter the objective to accomplish the mission. With less people involved, fewer can get hurt.

16 The Urban Survival Drill

- Where will most combat occur in the future?
- Why is that?

Squads members about to take turns moving up the street.

(Source: Courtesy of Sorman Information and Media, illustration by Wolfgang Bartsch, from "SoldF: Soldaten i falt," © 2001 Forsvarsmakten, Stockholm, pp. 22, 23.)

Safely Changing One's Urban Position Takes Rehearsal

Most future combat will occur in built-up areas, because their resident populations limit the use of heavy weaponry. A seasoned Special Weapons and Tactics (SWAT) member has already dealt with a wide variety of street frontages, door latches, building interiors, and room obstacles. Few urban scenarios would catch him off guard. But, for his military counterpart, enough muscle memory to do well in the city takes continual practice at moving up one side of a street without getting shot. (See Figure 16.1.)

Because several concurrent threats are encountered while ad-

Figure 16.1: Any One-Sided Street Will Work
(Source: FM 90-10-1 [1982], p. 3-35.)

Figure 16.2: A Two-Sided City Street for Training More People
(Source: FM 90-10-1 (1982), p. B-9)

vancing through built-up terrain, each student must take responsibility for his own improvement. For the best results, he works on one or two threats at a time and then several in combination. All can be done while negotiating a makeshift "street movement" course. At the end of each run, he gets feedback on how well he did against simulated enemy fire. By then repeating the course, he automatically improves.

Easiest Way to Provide That Training

No "hard-to-reserve" base combat town is needed for such a circuit. One can be quickly set up while any base business area is temporarily closed. The prerequisite rehearsal for a proficient urban warrior then becomes part of his daily PT. Below are the construction and operating details of such a street-following "fun run." Others for moving along hallways or through rooms can be established the same way.

First, a pair of instructors rig a deserted city block like one in the current war zone. They each take 14 or so students with rubber rifles to either side of the street on one end of this block. (See Figure 16.2 and the chapter page artwork.) Their subsequent operations will be separately conducted. Each instructor deploys his aggressors to the other end of the block to provide casualty simulation from two of the six main sources. (Glance forward at Table 16.1.) He then tells his students to avoid any three-second sight picture as they move one at a time up the store fronts. Every student will get successive tries at three pairs of threats.

During the first evolution, one aggressor sits in the middle of the street at the end of the block to simulate long-range MG or sniper fire, while the other stands next to the store fronts to simulate wall-following bullets. Both will count three-second sight pictures against each runner. Then, the number of hits is relayed to him before being sent back—behind the buildings—to the original start point. Upon arrival, the instructor records their score and gives everyone a second chance.

Then, two aggressors on either side of the street create a third and fourth less apparent threat. The instructor tells his students to avoid fire from between the buildings and upper stories. Each student gets two tries.

Figure 16.3: Any Hole in the Wall Can Emit Bullets
(Source: Courtesy of Sorman Info. & Media, illustration by Wolfgang Bartsch, from "SoldF: Soldaten i falt," *p. 289*, © 2001 Forsvarsmakten and Wolfgang Bartsch.)

Figure 16.4: Sunken Doorways Are Particularly Risky
(Source: Courtesy of Sorman Info. & Media, illustration by Wolfgang Bartsch, from "SoldF: Soldaten i falt," p. 289, © 2001 Forsvarsmakten and Wolfgang Bartsch.)

Finally, those same two aggressors provide a fifth and sixth concurrent threat. The instructor advises his runners to elude fire from any wall aperture or between the buildings. (See Figures 16.3 through 16.6.) Again, all get the opportunity to improve their performance.

One or two days later, the same instructors deploy six aggressors apiece and work on four threats at once, then all six. For a particularly motivated group, other threats from Table 16.1 can be subsequently added. Only one man can run the gauntlet every three minutes, so each column should be no bigger than 14 people. For more students, the command should set up duplicate courses at other locations. Where there is no pool of aggressors available, each squad can provide its own. Similar packages can be easily designed for interior movement.

Figure 16.5: All Windows Must Be Passed Under
(Sources: FM 90-10-1 (1982), p. B-29; FM 3-4 (1985), p. 3-5)

Figure 16.6: Spaces Between Buildings Also Pose a Threat
(Sources: FM 7-8 (1984), p. B-8; FM 90-10-1 (1982), p. B-17)

THINGS TO AVOID WHILE MOVING UP A CITY STREET

A searching GI can be easily seen coming.

He can be shot from several different sources:

By bullets deflected along walls
By long-range MG at either end of the street
From upper stories to the front, side, or rear
From spaces between buildings
From open windows and doors
From tiny wall apertures (like air vents or drainage holes)
From cracks or bullet/shell holes in the walls
From sewer openings in the gutters
From other grates in the sidewalks
From manhole covers or something over them (like a junked car)
From parked vehicles
From trash cans or piles of refuse on the sidewalk

Table 16.1: Ways of Getting Shot in the City

Why the PLA Will No Longer Pose as Big a Threat

Only foot soldiers agile and skilled enough to move around a contemporary battlefield without any supporting-arms umbrella will be able to defeat the light infantrymen of a bottom-up Asian society.

As long as the Pentagon bases most battlefield success on technology, the entry-level training of its "ground pounders" will be substandard. Over-controlled people don't make very durable warriors.

To escape the overhead surveillance and precision targeting of modern warfare, most future conflicts will occur in heavily populated built-up areas. As in Vietnam, China or one of its proxies will conduct mobile warfare resembling an insurgency, and only dismounted U.S. fireteams and buddy teams will be able to handle all the compartmentalization. U.S. tanks, planes, and artillery will be of little use. So, how proficient GIs can now become at dodging bullets will greatly effect the MW mission after emerging from their highly vulnerable personnel carriers.

Notes

SOURCE NOTES

Illustrations:

Cover art is in the public domain from WIKIMEDIA COMMONS. The file is named "Those_Who_Armed_Themselves_with_Mao_Zedong_Thought_Have_the_Greatest_Fighting_Strength,_China,_1970_-_Jordan_Schnitzer_Museum_of_Art,_University_of_Oregon_-_Eugene,_Oregon_-_DSC09548.jpg." It has been reproduced under a Creative Commons CCO 1.0 Universal Public Domain Dedication.

Image on page 9 is U.S. GOVERNMENT drawn or commissioned from http://search.usa.gov. See each caption for exact url.

Images on pages 15, 107, and 110 reprinted with permission of SORMAN INFORMATION AND MEDIA and SWEDISH ARMED FORCES, from *Soldf: Soldaten i Falt,* by Forsvarsmakten, with illustrations by Wolfgang Bartsch. These illustrations are from pages 22, 23, and 289 of the Swedish publication. Copyright © 2001 by Forsvarsmakten, Stockholm. All rights reserved.

Image on page 37 reprinted from "A CONCISE HISTORY OF THE UNITED STATES MARINE CORPS: 1775-1969." By Capt. William D. Parker. History Division, Headquarters Marine Corps, 1970. Sketch on page 62 by Capt. Donald L. Dickson.

Picture on page 41 reprinted from ARTPHOTO ARCHIVES of the U.S. Army's Center of Military History. Probably painted by an active-duty member, it is considered to be in the public domain with the following illustration designator: "avop08-01.2.jpg."

Picture on page 53 is reproduced after asking the permission of OSPREY. It is from "Armies of the Vietnam War 1962-75," *Men-at-Arms Series.* Plate C, Number 2. Copyright © 1980 by Osprey. All rights reserved.

Image on page 71 reprinted from "CLOSING IN: MARINES IN THE SEIZURE OF IWO JIMA," by Colonel Joseph H. Alexander, *Marines in WWII Commemorative Series,* Headquarters Marine Corps, 1994.

Picture on page 88 is from PUBLICDOMAINVECTORS.ORG with image designator "nicubunu_RPG_map_symbols_Townhall.jpg." For use by an educational nonprofit, it is considered to be in the public domain.

ENDNOTES

Introduction

1. Gen. Vo Nguyen Giap, as quoted in "The Tiger and the Elephant: Viet Minh Strategy and Tactics."
2. David M. Toczek, *The Battle of Ap Bac, Vietnam* (Westport, CT: Greenwood Press, 2001), pp. 50, 51.

Acknowledgments

1. Pope John Paul II, *Crossing the Threshold of Hope* (New York: Alfred A. Knopf, 1995), pp. 205, 206.

Chapter 1: *The Modern-Day Myth*

1. *Sinoland: The Subversion of Freedom's Bastion,* by H. John Poole (Emerald Isle, NC: Posterity Press, 2016); *Unrestricted Warfare,* by Qiao Liang and Wang Xiangsui (Beijing: PLA Literature and Arts Publishing House, February 1999), FBIS translation over the internet.
2. "Border Patrol says number of Chinese nationals entering U.S. has spiked almost 1,000%," *Fox News* on line, 20 March 2023.
3. "Warfighting," *MCDP 1* (Washington, D.C.: HQMC, 1997), pp. 42, 43.

Chapter 2: *How Much Does the New Gear Change Things?*

1. "Finnish Soldiers: Ukraine Needs Better Training to Win," by Mika Makelainen, *YLE News*, n.d.
2. *Guardian Joe,* by H. John Poole (Emerald Isle, NC: Posterity Press, 2018), chapt. 9, p. 81.
3. "Finnish Soldiers," by Makelainen.
4. Memorandum for the record from H.J. Poole, after viewing the video on ABC or PBS News around 1 June 2023, and then hearing about the lost personnel carriers from this or another equally reliable source.
5. "25 Tanks and Fighting Vehicles, Gone in a Blink: The Ukrainian Defeat near Mala Tokmachka Was Worst Than We Thought," by David Axe, *Forbes Business Aerospace & Defense,* 27 June 2023; "Nearly a third of the Bradley armored vehicles the US provided Ukraine may have already been lost or damaged," by Charles R. Davis, *Insider* on line, July 2023.

6. *The Iwo Alamo,* by H. John Poole (Emerald Isle, NC: Posterity Press, 2021), chapts. 10 and 11; *Okinawan Reckoning,* by H. John Poole (Emerald Isle, NC: Posterity Press, 2022), chapts. 10-15.

Chapter 3: *Single-File Travel*

1. *Advanced Tactics in America,* by H. John Poole (Emerald Isle, NC: Posterity Press, 2023), chapt. 10.

2. Burke Davis, *Marine* (New York: Bantam, 1964), p. 140; *Fighting on Guadalcanal,* FMFRP 12-110 (Washington, D.C.: U.S.A. War Office, 1942), pp. 2, 22; *Handbook on Japanese Military Forces,* TM 30-480 (Washington, D.C.: U.S. War Dept., 1944), p. 98.

3. *Phantom Soldier: The Enemy's Answer to U.S. Firepower,* by H. John Poole (Emerald Isle, NC: Posterity Press, 2001), chapt. 7.

4. Lynn Montross and Capt. Nicholas A. Canzona, "The Chosin Reservoir Campaign," *U.S. Marine Operations in Korea 1950-1953,* vol. III (Washington, D.C.: Hist. Branch, HQMC, 1957), p. 161.

5. Eric M. Hammel, *Chosin: Heroic Ordeal of the Korean War* (Novato, CA: Presidio Press, 1981), p. 87.

6. Montross and Canzone, *The Chosin Reservoir Campaign,* p. 163.

7. "Platoon," movie, written and directed by Oliver Stone, Hemdale Film Corporation, 1986.

8. Memorandum for the record by H.J. Poole, while Hdqts. Cmdt. of 27th Marines in Vietnam.

9. *Advanced Tactics in America,* by Poole, chapt. 5.

10. *Night Movements*, trans. and preface by C. Burnett (Tokyo: Imperial Japanese Army, 1913; reprint Port Townsend, WA: Loompanics Unlimited, n.d.), p. 40.

11. "Warfighting," MCDP 1, p.37.

12. Lt.Col. Whitman S. Bartley, *Iwo Jima: Amphibious Epic* (Washington, D.C.: Hist. Branch, HQMC, 1954), pp. 108, 109.

13. *Okinawa: The Last Battle,* by Roy E. Appleman, James M. Burns, Russell A. Gugeler, and John Stevens, *United States Army in World War II Series* (Washington, D.C.: U.S. Army Ctr. of Mil. Hist., 2000), p. 194.

Chapter 4: *What Requirement for Flankers?*

1. David M. Glantz, "August Storm: Soviet Tactical and Operational Combat in Manchuria, 1945," *Leavenworth Papers No. 8* (Ft. Leavenworth, KS: Combat Studies Inst., U.S. Army Cmd. & Gen. Staff College, 1983), p. 1.

2. "Historical Study—Night Combat," *DA Pamphlet 20-236* (Washington, D.C.: Hdqts. Dept. of the Army, 1953), pp. 19-21.

3. *The Tiger's Way: A U.S. Private's Best Chance for Survival,* by H. John Poole (Emerald Isle, NC: Posterity Press, 2003), pp. 79-85.

4. Maj. Scott R. McMichael, "The Chinese Communist Forces in Korea," chapt. 2 of "A Historical Perspective on Light Infantry," in *Leavenworth Research Survey No. 6* (Ft. Leavenworth, KS: Combat Studies Inst., U.S. Army Cmd. & Gen. Staff College, 1987), p. 58.

5. Ibid., pp. 56, 57.

6. *Night Movements,* trans. and preface by Burnett, pp. 79-82.

7. McMichael, "The Chinese Communist Forces in Korea," in *Leavenworth Research Survey No. 6,* p. 60.

8. Memorandum for the record from H.J. Poole, while CO of G/2/5 in Vietnam.

9. McMichael, "The Chinese Communist Forces in Korea," *in Leavenworth Research Survey No. 6,* p. 59.

10. Col. Joseph H. Alexander, "Battle of the Barricades: U.S. Marines in the Recapture of Seoul," *Marines in the Korean War Commemorative Series* (Washington, D.C.: Hist. and Museums Div., HQMC, 2000), pp. 8-10.

11. *The Strategic Advantage: Sun Zi and Western Approaches to War,* ed. Cao Shan (Beijing: New World Press, 1997), pp. 21, 55.

Chapter 5: *Dodging Foe Outposts and Roving Sentries*

1. David Scott-Donelan, *Tactical Tracking Operations: The Essential Guide for Military and Police Trackers* (Boulder, CO: Paladin Press, 1998), p. 24.

2. *German Squad Tactics in WWII,* by Matthew Gajkowski (West Chester,OH: Nafziger, 1995), reprint of *The German Squad in Combat,* trans. and ed. by the U.S. Mil. Intel. Service (N.p., 1943), p. 42.

3. *Night Movements*, trans. and preface by Burnett, p. 66.

4. Ibid., pp. 67, 68.

5. Edward J. Drea, "Nomonhan: Japanese—Soviet Tactical Combat, 1939," *Leavenworth Papers No. 2* (Ft. Leavenworth, KS: Combat Studies Inst., U.S. Army Cmd. & Gen. Staff College, 1981), p. 62.

6. *Night Movements*, trans. and preface by Burnett, pp. 70, 71.

7. Ibid., pp. 72, 73.

8. Ibid., p. 73.

9. *The Bear Went over the Mountain: Soviet Combat Tactics in Afghanistan,* trans. and ed. Lester W. Grau, Foreign Mil. Studies Office, U.S. Dept. of Defense (originally from the Soviet Union: Frunze Mil. Academy, n.d.; reprint Washington, D.C.: Nat. Defense Univ. Press, 1996), p. 129.

10. S.Sgt. Lee Bergee and PFC Fred Davidson, as quoted in *The Korean War: Pusan to Chosin, An Oral History,* by Donald Knox (San Diego: Harcourt Brace Jovanovich, 1985), pp. 289-291; *The Tiger's Way,* by Poole, p. 291.

Chapter 6: *The Elsewhere Located Base of Fire*

1. *Soviet Infantry Tactics in World War II: Red Army Infantry Tactics from Squad to Rifle Company from the Combat Regulations,* with trans., intro., and notes by Charles C. Sharp (West Chester, OH: Nafziger, 1998); first published as *Soviet Combat Regulations of November 1942* (Moscow: [Stalin], 1942), p. 21; *German Squad Tactics in WWII, by* Gajkowski, p. 15.
2. *German Squad Tactics in WWII,* by Gajkowski, p. 5.
3. "How a German Squad Attacks a Position (WWII)," *Footage Archive,* from a training film entitled "The German Infantry Squad in Action: A Demonstration of Minor Field Tactics."
4. Ibid.; *German Squad Tactics in WWII,* by Gajkowski p. 20.
5. *Soviet Infantry Tactics in World War II,* with trans., intro., and notes by Sharp, p. 19.
6. Ibid., pp. 18, 21.
7. Ibid., p. 19.
8. Ibid. p. 20.
9. "Saving Private Ryan," DVD, 169 minutes, Paramount Pictures, n.d.
10. Memorandum for the record from H.J. Poole, while on Operation Maui Peak in 1968 in Vietnam.

Chapter 7: *Are Stand-Up Assaults Absolutely Necessary?*

1. *German Squad Tactics in WWII,* by Gajkowski, p. 14.
2. "Iwo Jima: The Most Horrific of All World War II Battles," from Warfare Hist. Network, *The National Interest,* 15 December 2018; "Western Pacific Operations," by George W. Garand and Truman R. Strowbridge, vol. IV, part VI, chapt. 4, *History of Marine Corps Operations in World War II* (N.p: Hist. Branch, G-3 Div., HQMC, 1971), pp. 526, 527.
3. Hammel, *Chosin,* p. 87.
4. Joseph C. Goulden, *Korea: The Untold Story of the War* (New York: Times Books, 1982), p. 355.
5. Ibid., p. 350.
6. Hammel, *Chosin,* pp. 79, 80.
7. *Night Movements*, trans. and preface by Burnett, p. 122.

8. *Handbook on Japanese Military Forces,* TM 30-480 (Washington, D.C.: U.S. War Dept., 1944), p. 98.
9. Montross and Canzone, *The Chosin Reservoir Campaign,* p. 163.
10. *Handbook of the Chinese People's Liberation Army,* DDB-2680-32-84 (Washington, D.C.: DIA, November 1984), pp. 16, A-5, A-15.
11. Montross and Canzone, *The Chosin Reservoir Campaign,* p. 172.
12. "How Did the Chinese Army Defeat the U.S. Tanks and Helicopters During the Korean War," photograph, *Quora,* n.d.
13. Hammel, *Chosin,* p. 87.
14. Hammel, *Chosin,* pp. 55, 155, 156.
15. Montross and Canzone, *The Chosin Reservoir Campaign,* p. 163.
16. Hammel, *Chosin,* p. 89.
17. Montross and Canzone, *The Chosin Reservoir Campaign,* p. 163.
18. Hammel, *Chosin,* p. 87.
19. Ibid., p. 55.
20. *East of Chosin: Entrapment and Breakout in Korea, 1950,* by Roy Edgar Appleman (College Station, TX: Texas A&M Univ. Press, 1990), p. 125.
21. Montross and Canzone, *The Chosin Reservoir Campaign,* p. 168.
22. McMichael, "The Chinese Communist Forces in Korea," *Leavenworth Research Survey No. 6,* p. 62.
23. "Handbook on Japanese Military Forces," *TM-E 30-480,* p. 86.
24. Hammel, *Chosin,* pp. 89, 90.
25. Ibid., p. 68
26. *Fighting on Guadalcanal,* FMFRP 12-110 (Washington, D.C.: U.S.A. War Office, 1942), p. 22.
27. Hammel, *Chosin,* p. 59.
28. Ibid., pp. 56-59; Don Campbell (Marine platoon sergeant during the bloody hill battles of 1952-53), telephone conversation with author, 26 March 2001.
29. McMichael, "The Chinese Communist Forces in Korea," *Leavenworth Research Survey No. 6,* p. 63.
30. Hammel, *Chosin,* pp. 71, 72.

Chapter 8: *Gun Bunker Takedowns*

1. *Peleliu Progress,* by H. John Poole (Emerald Isle, NC: Posterity Press, 2021), chapt. 5.
2. Hammel, *Chosin,* p. 87.
3. *The Strategic Advantage,* ed. Cao Shan, pp. 2, 64.

4. *The Way of the Warrior,* editors of Time-Life Books (Alexandria, VA: Time-Life Books, 1993), pp. 129-131, in *Frontier Skills: The Tactics and Weapons That Won the American West,* by *Frontier,* by William C. Davis (Guilford, CT: Lyons Press, 2003), p. 15.

5. "Platoon," movie, written and directed by Oliver Stone.

6. "How a German Squad Attacks a Position (WWII)," *Footage Archive,* from a training film entitled "The German Infantry Squad in Action."

7. *Robert Rogers' Rules for the Ranging Service: An Analysis,* by Matt Wulff (Westminster, MD: Heritage Books, 2007), p. 83.

8. "We Were Soldiers Once and Young," DVD, 138 minutes, directed by Randall Wallace, Paramount, n.d.

9. *Wikipedia Encyclopedia*, s.v. "RPG-7" and "M153 SMAW."

10. *Okinawan Reckoning,* by H. John Poole (Emerald Isle, NC: Posterity Press, 2022), p. 200; "The Final Campaign: Marines in the Victory on Okinawa," by Alexander, NPS version (Closing the Loop), DoD Photo (USMC) 122167.

Chapter 9: *With Regard to Gaps and Firesacks*

1. "The Final Campaign: Marines in the Victory on Okinawa," by Col. Joseph Alexander, *Marines in World War II Commemorative Series* (Washington, D.C.: Hist. and Museums Div., HQMC, 1996), p. 20.

2. *Okinawa: The Last Battle,* by Appleman et al, p. 321.

Chapter 10: *Objective Consolidation*

1. Memorandum for the record from H.J. Poole, after hearing a Falkland Island newscast around 1982.

2. H.J. Poole, *The Last Hundred Yards: The NCO's Contribution to Warfare* (Emerald Isle, NC: Posterity Press, 1997), p. 350.

Chapter 11: *With Common Sense as One's Compass*

1. FMFM 6-4, *Marine Rifle Company/Platoon,* Marine Corps Develop. & Educ. Cmd. (Washington, D.C.: HQMC, 1978), p. 5.

2. "Gung Ho!: The Story of Carlson's Makin Island Raiders," movie, 87 minutes, Universal Studios, 1943.

3. Col. Merritt A. Edson, as quoted in *Fighting on Guadalcanal,* FMFRP 12-110, pp. 14-19.

4. "From Makin to Bougainville: Marine Raiders in the Pacific War," by Maj. Jon T. Hoffman, *Marines in WWII Commemorative Series* (Washington, D.C.: Marine Corps Hist. Ctr., 1995), p. 23.

5. "Gung Ho," movie, 87 minutes, Universal Studios.

Chapter 12: *Grenade Throwing for Distance & Accuracy*

1. *The Iwo Alamo,* by Poole, chapt. 2; *Pelelilìu Progress,* by Poole, chapt 7.

2. *Peleliu Progress,* by Poole, chapt. 5.

3. *The Iwo Alamo,* by Poole, chapt. 3.

4. "Sands of Iwo Jima," videocassette, 109 minutes, Republic Pictures, 1988.

5. "Jääkärijoukkueen ja-Ryhmä Käsikirja," Rifle-Light Infantry Platoon and Squad Handbook, Finnish Army HN878, trans. John Manter (Helsinki, 2018), p. 127.

Chapter 13: *Crawling for Distance and Secrecy*

1. Sgt. O.J. Marion, as quoted in *Fighting on Guadalcanal,* FMFRP 12-110 (Washington, D.C.: U.S.A. War Office, 1942), pp. 25, 26; *Handbook on Japanese Military Forces,* TM 30-480, p. 98.

Chapter 14: *Daily Boots and Rifles Fun Run*

1. "Tactics," *MCDP 1-3* (Washington, D.C.: HQMC, 1997), p. 62.

2. Bruce I. Gudmundsson, *Stormtroop Tactics—Innovation in the German Army 1914-1918* (New York: Praeger, 1989), pp. 92-94, 167-168.

3. Davis, *Marine*, p. 140; *Fighting on Guadalcanal,* FMFRP 12-110, pp. 2, 22.

4. Hammel, *Chosin*, p. 87.

5. "Platoon," movie, written and directed by Stone.

Chapter 15: *Stalking and Short-Range Infiltration Course*

1. Michael Lee Lanning and Dan Cragg, *Inside the VC and the NVA: The Real Story of North Vietnam's Armed Forces* (New York: Ivy Books, 1992), pp. 211, 212.

GLOSSARY

ARVN	Army of the Republic of Vietnam	South Vietnamese soldiers.
BOF	Base of Fire	Bullets to keep the defender heads down during a U.S. assault
CAP	Combined Action Platoon	Village defense unit comprised of three local militia squads, each having one U.S. Marine fireteam
CCF	Chinese Communist Forces	Foe in Korean War
CP	Command Post	Unit headquarters
DMZ	Demilitarized Zone	Boundary between the two Vietnams.
DoD	Department of Defense	Pentagon agency
4GW	4th Generation Warfare	Four-arena conflict: martial, political, economic, religious/ psychological
FMFM	Fleet Marine Corps Field Manual	Official USMC publication
FPL	Final Protective Line	Interlocking MG fire
GI	Government Issue	Colloquial term for low-ranking member of U.S. Armed Forces

GPS	Global Positioning System	Device that resects satellite signals to determine location.
LP	Listening Post	Nighttime observation post
MCDP	Marine Corps Doctrinal Publication	USMC manual
MG	Machinegun	Automatic weapon
MILES	Multiple Integrated Laser Engagement System	Casualty simulation device
MOS	Military Occupational Specialty	Career field
MW	Maneuver Warfare	Way of fighting where surprise takes place of firepower.
NCO	Non-Commissioned Officer	Junior enlisted military leader.
NKPA	North Korean People's Army	Pyongyang's ground forces
NVA	North Vietnamese Army	Hanoi's ground forces
PLA	People's Liberation Army	Beijing's military forces
PRC	People's Republic of China	Communist China
PT	Physical Training	Calisthenics being practiced.
RPG	Rocket Propelled Grenade	Ordnance launched by Communist bazooka-like weapon
2GW	2nd Generation Warfare	Focus is on killing foe soldiers; same as traditional U.S. style (attrition warfare).

SMAW	Shoulder-Launched Multipurpose Assault Weapon	Hand-held bunker busting missile launcher
SNCO	Staff Non-Commissioned Officer	Senior enlisted military leader.
SWAT	Special Weapons and Tactics	Police unit skilled at urban assault
TAOR	Tactical Area of Responsibility	Zone of action.
3GW	3rd Generation Warfare	Avoids foe defenders to more easily get at command and logistic hubs.
U.N.	United Nations	Worldwide alliance of countries.
U.S.	United States	America
USMC	United States Marine Corps	America's amphibious force in readiness.
UW	Unconventional Warfare	Attacking through infiltration and defending through ambushes; part of 3GW.
WWI	World War I	First global conflict.
WWII	World War II	Second global conflict.
WWIII	World War III	Third global conflict.

Bibliography

U.S. Government Publications, News Releases, and Videos

Alexander, Col. Joseph H. "Battle of the Barricades: U.S. Marines in the Recapture of Seoul." *Marines in the Korean War Commemorative Series*. Washington, D.C.: History and Museums Division, Headquarters Marine Corps, 2000.

Bartley, Lt.Col. Whitman S. *Iwo Jima: Amphibious Epic*. Washington, D.C.: History Branch, Headquarters Marine Corps, 1954.

The Bear Went over the Mountain: Soviet Combat Tactics in Afghanistan. Translated and edited by Lester W. Grau. Foreign Military Studies Office, U.S. Department of Defense. Originally from the Soviet Union: Frunze Military Academy, n.d. Reprint Washington, D.C.: National Defense University Press, 1996.

Drea, Edward J. "Nomonhan: Japanese—Soviet Tactical Combat, 1939." *Leavenworth Papers No. 2*. Fort Leavenworth, KS: Combat Studies Institute, U.S. Army Command and General Staff College, 1981.

Fighting on Guadalcanal. FMFRP 12-110. Washington, D.C.: U.S.A. War Office, 1942.

"The Final Campaign: Marines in the Victory on Okinawa." By Col. Joseph Alexander. *Marines in World War II Commemorative Series*. Washington, D.C.: History and Museums Division, Headquarters Marine Corps, 1996.

FMFM 6-4. *Marine Rifle Company / Platoon*. Marine Corps Development and Education Command. Washington, D.C.: Headquarters U.S. Marine Corps, 1978.

FMFRP 12-110. *Fighting on Guadalcanal*. Washington, D.C.: U.S.A. War Office, 1942.

"From Makin to Bougainville: Marine Raiders in the Pacific War." By Maj. Jon T. Hoffman. *Marines in WWII Commemorative Series*. Washington, D.C.: Marine Corps History Center, 1995.

German Squad Tactics in WWII. By Matthew Gajkowski. West Chester, OH: Nafziger, 1995. Reprint of *The German Squad in Combat.* Translated and edited by the U.S. Military Intelligence Service. N.p., 1943.

Glantz, David M. "August Storm: Soviet Tactical and Operational Combat in Manchuria, 1945." *Leavenworth Papers No. 8.* Fort Leavenworth, KS: Combat Studies Institute, U.S. Army Command and General Staff College, 1983.

Handbook of the Chinese People's Liberation Army. DDB-2680-32-84. Washington, D.C.: Defense Intelligence Agency, November 1984. As retrieved in December 2018 from the following url: https://archive.org/stream/pdfy-WidHRLGeFX0jiTJF/Handbook+Of+The+Chinese+People%27s+Liberation+Army_djvu.txt.

Handbook on Japanese Military Forces. TM 30-480. Washington, D.C.: U.S. War Department, 1944. Reprint. Baton Rouge, LA: Louisiana State University Press, 1991.

"Historical Study—Night Combat." *DA Pamphlet 20-236.* Washington, D.C.: Headquarters Department of the Army, 1953.

McMichael, Maj. Scott R. "The Chinese Communist Forces in Korea." Chapter 2 of "A Historical Perspective on Light Infantry." *Leavenworth Research Survey No. 6.* Fort Leavenworth, KS: Combat Studies Institute, U.S. Army Command and General Staff College, 1987.

Montross, Lynn and Capt. Nicholas A. Canzona. "The Chosin Reservoir Campaign." *U.S. Marine Operations in Korea 1950-1953.* Volume III. Washington, D.C.: History Branch, Headquarters Marine Corps, 1957.

Okinawa: The Last Battle. By Roy E. Appleman, James M. Burns, Russell A. Gugeler, and John Stevens. *United States Army in World War II Series.* Washington, D.C.: U.S. Army Center of Military History, 2000.

"Tactics." M*CDP 1-3.* Washington, D.C.: Headquarters Marine Corps, 1997.

"Warfighting." MCDP 1. Washington, D.C.: Headquarters Marine Corps, 1997.

"Western Pacific Operations." By George W. Garand and Truman R. Strowbridge. Volume IV, part VI, chapters 1 and 4. *History of Marine Corps Operations in World War II.* N.p.: Historical Branch, G-3 Division, Headquarters Marine Corps, 1971. As retrieved in January 2020 from the following urls: https://www.ibiblio.org/hyperwar/USMC/IV/USMC-IV-VI-1.html and https://www.ibiblio.org/hyperwar/USMC/IV/USMC-IV-VI-4.html. And chapter 11. As retrieved from this url: https://www. ibiblio.org/hyperwar/USMC/IV/USMC-IV-VI-11.html. Whole book available from this url: https://www.ibiblio.org/hyperwar/USMC/IV/index.html.

Civilian Publications

<u>*Analytical Studies, Websites, and Personal Conversations*</u>

Advanced Tactics in America. By H. John Poole. Emerald Isle, NC: Posterity Press, 2023.

Campbell, Don (Marine platoon sergeant during the bloody hill battles of 1952-53). Telephone conversation with author, 26 March 2001.

Davis, Burke. *Marine.* New York: Bantam, 1964.

East of Chosin: Entrapment and Breakout in Korea, 1950. By Roy Edgar Appleman. College Station, TX: Texas A&M University Press, 1990.

Frontier Skills: The Tactics and Weapons That Won the American West. By William C. Davis. Guilford, CT: Lyons Press, 2003.

Goulden, Joseph C. *Korea: The Untold Story of the War.* New York: Times Books, 1982.

Guardian Joe. By H. John Poole. Emerald Isle, NC: Posterity Press, 2018.

Gudmundsson, Bruce I. *Stormtroop Tactics—Innovation in the German Army 1914-1918.* New York: Praeger, 1989.

Hammel, Eric M. *Chosin: Heroic Ordeal of the Korean War.* Novato, CA: Presidio Press, 1981.

The Iwo Alamo. By H. John Poole. Emerald Isle, NC: Posterity Press, 2021.

"Jääkärijoukkueen ja-Ryhmä Käsikirja." Rifle-Light Infantry Platoon and Squad Handbook. Finnish Army HN878. Translated by John Manter. Helsinki, 2018.

John Paul II, Pope. *Crossing the Threshold of Hope.* New York: Alfred A. Knopf, 1995.

The Korean War: Pusan to Chosin, An Oral History. By Donald Knox. San Diego: Harcourt Brace Jovanovich, 1985.

Lanning, Michael Lee and Dan Cragg. *Inside the VC and the NVA: The Real Story of North Vietnam's Armed Forces.* New York: Ivy Books, 1992.

Night Movements. Translated and preface by C. Burnett. Tokyo: Imperial Japanese Army, 1913. Reprint Port Townsend, WA: Loompanics Unlimited, n.d.

Okinawan Reckoning. By H. John Poole. Emerald Isle, NC: Posterity Press, 2022.

Peleliu Progress. By H. John Poole. Emerald Isle, NC: Posterity Press, 2021.

Phantom Soldier: The Enemy's Answer to U.S. Firepower. By H. John Poole. Emerald Isle, NC: Posterity Press, 2001.

Poole, H.J. *The Last Hundred Yards: The NCO's Contribution to Warfare.* Emerald Isle, NC: Posterity Press, 1997.

Robert Rogers' Rules for the Ranging Service: An Analysis. By Matt Wulff. Westminster, MD: Heritage Books, 2007.

Scott-Donelan, David. *Tactical Tracking Operations: The Essential Guide for Military and Police Trackers.* Boulder, CO: Paladin Press, 1998.

Sinoland: The Subversion of Freedoms Bastion. By H. John Poole. Emerald Isle, NC: Posterity Press, 2016.

Soviet Infantry Tactics in World War II: Red Army Infantry Tactics from Squad to Rifle Company from the Combat Regulations. With translation, introduction, and notes by Charles C. Sharp. West Chester, OH: Nafziger, 1998. First published as *Soviet Combat Regulations of November 1942.* Moscow: [Stalin], 1942.

The Strategic Advantage: Sun Zi & Western Approaches to War. Edited by Cao Shan. Beijing: New World Press.

The Tiger's Way: A U.S. Private's Best Chance for Survival. By H. John Poole. Emerald Isle, NC: Posterity Press, 2003.

Toczek, David M. *The Battle of Ap Bac, Vietnam.* Westport, CT: Greenwood Press, 2001.

Unrestricted Warfare. By Qiao Liang and Wang Xiangsui. Beijing: PLA Literature and Arts Publishing House, February 1999. FBIS translation over the internet.

Wikipedia Encyclopedia. From its website, https://en.wikipedia.org.

Videotapes, Movies, TV Programs, Slide Shows, DVDs, CDs, Illustrations

"The Battle of Biak: 'A Terrifying Glimpse Into the Soul of Mankind.'" By David Alan Johnson. *Warfare History Network,* n.d. As retrieved in September 2020 from the following url: https://warfarehistorynetwork.com/2018/.

"The Final Campaign: Marines in the Victory on Okinawa." By Colonel Joseph H. Alexander. NPS version (Closing the Loop). DoD Photo (USMC) 122167.

"Gung Ho!: The Story of Carlson's Makin Island Raiders." Movie, 87 minutes. Universal Studios, 1943.

"How a German Squad Attacks a Position (WWII)." *Footage Archive.* From a training film entitled "The German Infantry Squad in Action: A Demonstration of Minor Field Tactics." As retrieved in December 2018 from this url: www,youtube,com/watch?v=GDZMJXaADQI.

"How Did the Chinese Army Defeat the U.S. Tanks and Helicopters During the Korean War." Photograph. *Quora,* n.d. As retrieved in March 2019 from the following url: https://www.quora.com/How-did-the-Chinese-army-defeat-the-US-tanks-and-helicopters-during-the-Korean-War.

"Platoon." Movie. Written and directed by Oliver Stone. Hemdale Film Corporation, 1986.

"Sands of Iwo Jima," Videocassette, 109 minutes. Republic Pictures, 1988.

"Saving Private Ryan." DVD, 169 minutes. Paramount Pictures, n.d.

"We Were Soldiers Once, and Young." DVD, 138 minutes. Directed by Randall Wallace. Paramount, n.d.

Newspaper, Magazine, and Website Articles

"Border Patrol says number of Chinese nationals entering U.S. has spiked almost 1,000%." *Fox News* on line, 20 March 2023.

"Finnish Soldiers: Ukraine Needs Better Training to Win." By Mika Makelainen. *YLE News*, n.d. As retrieved in April 2023 from the following url: https://yle.fi/a/74-20020197.

Giap, Gen. Vo Nguyen. As quoted in "The Tiger and the Elephant: Viet Minh Tactcs and Strategy." As retrieved in June 2023 from the following url: http://indochine54.free.fr/vm/tiger.html.

"Iwo Jima: The Most Horrific of All World War II Battles." By Warfare History Network. *The National Interest,* 15 December 2018. As retrieved in January 2020 from the following url: https://nationalinterest.org/blog/buzz/iwo-jima-most-horrific-all-world-war-ii-battles-38897?page=0%2C1.

"Nearly a third of the Bradley armored vehicles the US provided Ukraine may have already been lost or damaged." By Charles R. Davis. *Insider* on line, July 2023. As retrieved in July 2023 from the following url: https://www.businessinsider.com/ukraine-nearly-third-of-bradley-armored-vehicles-lost-or-damaged-2023-7.

"25 Tanks and Fighting Vehicles, Gone in a Blink: The Ukrainian Defeat near Mala Tokmachka Was Worst Than We Thought." By David Axe. *Forbes Business Aerospace & Defense,* 27 June 2023. As retrieved on 30 June 2023 from the following url: https://www.forbes.com/sites/davidaxe/2023/06/27/25-tanks-and-fighting-vehicles-gone-in-a-blink-the-ukrainian-defeat-near-mala-tokmachka-was-worst-than-we-thought/.

About the Author

After 28 years of commissioned and noncommissioned infantry service, John Poole retired from the United States Marine Corps in April 1993. While on active duty, he studied small-unit tactics for nine years: (1) six months at the Basic School in Quantico (1966); (2) seven months as a rifle platoon commander in Vietnam (1966-67); (3) three months as a rifle company commander at Camp Pendleton (1967); (4) five months as a regimental headquarters company (and camp) commander in Vietnam (1968); (5) eight months as a rifle company commander in Vietnam (1968-69); (6) five and a half years as an instructor with the Advanced Infantry Training Company (AITC) at Camp Lejeune (1986-92); and (7) one year as the Staff Noncommissioned Officer in Charge (SNCOIC) of the 3rd Marine Division Combat Squad Leaders Course (CSLC) on Okinawa (1992-93).

While at AITC, he developed, taught, and refined courses on maneuver warfare, land navigation, fire support coordination, call for fire, adjust fire, close air support, M203 grenade launcher, movement to contact, daylight attack, night attack, infiltration, defense, offensive Military Operations in Urban Terrain (MOUT), defensive MOUT, Nuclear/Biological/Chemical (NBC) defense, and leadership. While at CSLC, he further refined the same periods of instruction and developed others on patrolling.

He has completed all of the correspondence school requirements for the Marine Corps Command and Staff College, Naval War College (1,000-hour curriculum), and Marine Corps Warfighting Skills Program. He is a graduate of the Camp Lejeune Instructional Management Course, the 2nd Marine Division Skill Leaders in Advanced Marksmanship (SLAM) Course, and the East-Coast School of Infantry Platoon Sergeants' Course.

In the 25 years since retirement, Poole has researched the small-unit tactics of other nations and written 18 other books: (1) *The Last Hundred Yards,* a squad combat study based on the consensus opinions of 1,200 formal school NCOs and casualty statistics of AITC and CSLC field trials; (2) *One More Bridge to Cross,* a treatise on enemy proficiency at short range and how to match it; (3) *Phantom Soldier,* an in-depth look at the highly deceptive Asian style of war; (4) *The Tiger's Way,* the fighting styles of Eastern fireteams and soldiers; (5) *Tactics of the Crescent Moon,* insurgent procedures in Palestine, Chechnya, Afghanistan, and Iraq; (6) *Militant Tricks,* how Muslim insurgents fight in an urban environment; (7) *Terrorist Trail,* tracing the *jihadists* in Iraq back to their home countries; (8) *Dragon Days,* an unconventional warfare technique manual; (9) *Tequila Junction,* how to fight narco-guerrillas; (10) *Homeland Siege,* countering the 4GW assault on America by a foreign power's organized-crime proxies; (11) *Expeditionary Eagles,* how to outmaneuver the Taliban; (12) *Global Warrior,* forestalling WWIII through tiny-contingent activity; (13) *Gung Ho,* how supporting arms are not needed to take strongpoint matrices; (14) *Strategic Rifleman,* how to produce tiny semi-autonomous combat elements; (15) *Afrique,* with Africa as a model how U.S. intelligence is being gathered incorrectly for certain adversaries; (16) *Sinoland,* how the U.S. has been under 4GW attack from China since 9/11; (17) *Guardian Joe,* how—through more skilled infantry squads—the Pentagon can use less force to more easily win a modern war; (18) *North Korea's Hidden Assets,* 13 little-known wartime capabilities of what has been erroneously called the Hermit Kingdom; (19) *Super-Squad: The Now Missing Component;* (20) *The Iwo Alamo;* (21) *Peleliu Progress; (22) Okinawan Reckoning*; and (23) *Advanced Tactics in America.*

Since retirement, Poole has traveled extensively in both Communist and Islamist worlds to study the infantry tactics of America's enemies. He has done research in Mainland China (twice), North Korea, Vietnam, Cambodia, Pakistan (twice), Bangladesh, Malaysia, Iran, Lebanon, Turkey, the Emirates, Egypt, Sudan, Tanzania, Zambia, South Africa, Morocco, and Venezuela. Over the course of his lifetime, he has visited scores of other nations on all five continents. As of April 2014, John Poole had conducted multi-day training sessions (on advanced squad tactics) at 41 (mostly Marine) battalions, nine Marine schools, and seven special-operations units from all four U.S. service branches.

Between early tours in the Marine Corps (from 1969 to 1971), he served as a criminal investigator with the Illinois Bureau of Investigation (IBI). After attending the State Police Academy for several months in Springfield, he was assigned to the IBI's Chicago office. There, he worked mostly on general criminal and drug cases.

Several periods in Poole's life have qualified him to accurately mirror the viewpoints of America's true fighters— its junior enlisted infantrymen. In 1965, his first official orders were for "Private Poole." From 1992 to 1993, he served as SNCOIC of a divisional squad leaders school. Then, almost all of the 1100 students who attended his "post-retirement" training sessions were E-5's and below.

Poole is one of the very few Americans with an extensive enough background to know how to fix the ongoing tactical shortfall within the U.S. military. After some 60 multi-day training sessions at different active-duty units, he has developed, tested, and refined its solution (through a "bottom-up" training supplement). He has also become adept at the now virtually lost art of field training light infantrymen (those who are able do well on their own in the absence of orders).

Name Index

A

B

C

D

No entries.

E

F

No entries.

G

H

I

J

K

No entries.

L

No entries.

M

N

No entries.

O

No entries.

P

No entries.

Q

No entries.

R

S

T

U

V

No entries.

W

No entries.

X

No entries.

Y

No entries.

Z